Growing Fruits, Berries & Nuts
Southwest-Southeast

Second Edition

9 780884 150404

51295

Other books of interest from
Gulf Publishing Company

- Vegetable Growing for Southern Gardens

- Herb Gardening in Texas

- A Garden Book for Houston and the Gulf Coast

- Beachcomber's Guide to Gulf Coast Marine Life

- Hiking and Backpacking Trails of Texas

- Camper's Guide to Texas

- Camper's Guide to Florida

Growing Fruits, Berries & Nuts
Southwest-Southeast

George Ray McEachern

Second Edition

Gulf Publishing Company

Growing Fruits, Berries & Nuts
Southwest-Southeast

SECOND EDITION

First Edition, May 1978
Second Printing, May 1981

Second Edition, January 1990

Library of Congress Cataloging-in-Publication Data

McEachern, George Ray.
  Growing fruits, berries & nuts, Southwest-Southeast/George Ray McEachern.—2nd ed.
    p. cm.
  Rev. ed. of: Growing fruits, berries & nuts in the South. c1978.

  ISBN 0-88415-040-2

  1. Fruit-culture—Southern States. 2. Nuts—Southern States. I. McEachern, George Ray. Growing fruits, berries & nuts in the South. II. Title.
SB355.M16   1990   89-17138
634'.0975—dc20   CIP

Contents

To my teachers

Who showed me the fruit of the South,
Taught me how to grow it,
Challenged me to understand it, and
Demonstrated how to love it.

Murphy W. McEachern
County Agent
Plaquemines Parish, Louisiana

J. Benton Storey
Professor of Horticulture
Texas A&M University

P. Lynn Hawthorne
Professor of Horticulture
Louisiana State University

Bluefford G. Hancock
Extension Horticulturist
Texas A&M University

Acknowledgments

I would like to thank Earl Puls, David Creech, and Bobby Reeder for reviewing the manuscript and making important suggestions. Thanks also to Elizabeth Raven McQuinn, editor, and the staff at Gulf Publishing Company. Photographs were generously supplied by Benton Storey, Hollis Bowen, Jerry Parsons, Bill Welch, and Earl Puls. Appreciation is extended to Jerral Johnson and Charles Cole for use of their spray schedule. Steve Myers, Mike Smith, Justin Morris, Richard Mullenax, and Mike Kilby assisted with identifying the fruit producing regions.

To the fruit growers, county Extension agents, research horticulturists, Extension specialists, and secretaries who have helped me along the way, my sincere appreciation.

To my family: Jo Lynn, Charis, Shelly, Willie Dee, Arvella, and Donnie, who have always tolerated as well as inspired me, I will always be indebted.

Preface

This book was prepared for those interested in or actively growing fruit for a profit in the southwest to southeast regions of the United States. It is for the small commercial or family fruit grower. I have attempted to be simple, concise, straightforward, and as illustrative as possible so that beginners and professionals alike can use the information herein.

I hope most of all that my readers will recognize the great potential we southerners now have for growing our own fruits.

George Ray McEachern
College Station, Texas

Climate Data for Major Cities

	Last Spring Freeze	First Fall Freeze	Frost-Free Days	Record January Low (°F)	Minimum Hours of Chilling	Inches of Rain
Alabama						
Birmingham	Mar. 19	Nov.14	241	1	1,000	53
Huntsville	Apr. 1	Nov. 8	221	−9	1,100	50
Mobile	Feb. 17	Dec. 12	298	14	500	67
Montgomery ...	Feb. 27	Dec. 3	279	5	700	51
Arizona						
Phoenix	Feb. 14	Dec. 6	295	20	400	7
Tuscon	Mar. 15	Nov. 22	252	16	600	11
Arkansas						
Little Rock	Mar. 16	Nov. 15	244	−4	1,000	49
Florida						
Jacksonville	Feb. 6	Dec. 16	313	2	400	53
Orlando	Jan. 31	Dec. 17	319	24	300	51
Tampa	Jan. 10	Dec. 26	349	23	200	51
Georgia						
Atlanta	Mar. 20	Nov. 19	244	−3	800	49
Macon	Mar. 14	Nov. 7	240	3	700	44
Savannah	Feb. 21	Dec. 9	291	9	500	48
Kentucky						
Lexington	Apr. 13	Oct. 28	198	−15	1,400	43
Louisville	Apr. 1	Nov. 7	220	−20	1,400	41
Louisiana						
Baton Rouge ...	Feb. 28	Nov. 30	275	10	500	55
New Orleans ...	Feb. 13	Dec. 9	300	14	400	64
Maryland						
Baltimore	Mar. 26	Nov. 19	238	−7	1,400	43
Mississippi						
Jackson	Mar. 10	Nov. 13	248	7	700	50
New Mexico						
Albuquerque ...	Apr. 16	Oct. 29	196	−17	1,200	8
Roswell	Apr. 9	Nov. 2	208	−8	1,200	12
North Carolina						
Charlotte	Mar. 21	Nov. 15	239	4	900	43
Greensboro	Mar. 24	Nov. 16	237	0	1,100	43
Oklahoma						
Oklahoma City .	Mar. 28	Nov. 7	223	0	1,200	32
Tulsa	Mar. 31	Nov. 2	216	−2	1,300	38
South Carolina						
Charleston	Feb. 19	Dec. 10	294	11	600	49
Columbia	Mar. 14	Nov. 21	252	5	700	47
Tennessee						
Knoxville	Mar. 31	Nov. 6	220	−16	1,100	45
Memphis	Mar. 20	Nov. 12	237	−8	1,000	49
Nashville	Mar. 28	Nov. 7	224	−6	1,100	47
Texas						
Austin	Mar. 15	Nov. 20	244	12	700	33
Dallas-Ft. Worth	Mar. 18	Nov. 17	244	5	1,000	33
Houston	Feb. 10	Dec. 8	301	19	600	44
San Antonio ...	Feb. 24	Dec. 3	282	0	600	26
Virginia						
Norfolk	Apr. 4	Nov. 9	219	10	1,100	44
Richmond	Apr. 20	Oct. 18	181	−12	1,200	44

Major Commercial Fruit Producing Areas of the Southwest-Southeast

N

apples	citrus	strawberries
peaches	grapes	kiwifruit
pecans	muscadines	blueberries

What You Must Know to Begin

If you live in the southern United States, your opportunities for growing fruit are excellent. Our soils, climates, and adapted varieties permit growing a wide range of delicious fruit, and the enjoyment you get from tasting your own produce is one of life's greatest.

As one travels from Charleston, South Carolina to Tucson, Arizona, from Louisville, Kentucky to Mobile, Alabama, one recognizes beautiful orchards that lace the South. Large pecan trees shade our homes. Muscadines, blackberries, and strawberries border our lawns and trellis our patios. Yes, you can grow your own fruits and berries.

Millions of southerners are returning to the suburbs and small towns to enjoy country living, open air, a clean environment—the good earth. As they pioneer this new lifestyle, growing their own fruit becomes a natural, important part of family life.

Fruits add beauty to the landscape, delight to meals, and pleasure to work. Growing one's own fruit is a sig-

. . . and those that you can manage successfully in a limited space.

In this book you'll learn about fruits you can plant as a commercial or semi-commercial orchard . . .

nificant means of pride and enjoyment. The orchard is a place where the family can work together and earn money. Children can learn how plants grow, produce food, and multiply. Fruit culture can be profitable, even on a small scale. Adults escape from the rushed world to relax in the simple, healthy chores of culturing the trees.

Yet fruit culture is one of man's most sophisticated experiences with nature and the environment. Producing top-quality fruit fosters a real understanding of the environment and of horticulture.

Our wide range of southern crops offers opportunities for those with a small, limited space and for those with 10 acres or more. These different fruits can ripen essentially all year round.

1

What Do You Do First?

Check Out Your Soil. It can be excellent or terrible. Fortunately, most southern soils produce good fruit trees. Some soils may require special attention or specific crops, so read the section on soils (pages 8–10).

Study Your Climate. Our southern weather is very different from up north or way out west. Actually, it differs greatly within a single state. What is the total number of chilling hours in your area? How many frost-free days do you have? What is your humidity and rainfall situation? These are important questions which you can answer by reading the section on climate (pages 10–15).

Evaluate Your Site. Use every square foot to its full potential, but don't crowd your plants. One healthy oriental pear tree is much more attractive, productive, and rewarding in a 10 × 20-foot space than a row of blackberries, two peaches, two plums, and a pecan tree. Study your site carefully.

What Fruits Should You Plant?

After evaluating your soil, understanding your climate, and sizing up your site, you're ready to think about what you can grow. Visit neighbors and your county Extension agent to learn what is already growing well in your local soil and climate. Don't write grandma back in Michigan or your friend in California—their crops do not work in the South. Literature prepared for the U.S. in general or for the North or the West Coast is of little value when growing southern fruits. National mail order catalogs or newspaper ads can lead you in the wrong direction, so follow your local authorities' advice. Keep reading—this book is prepared specifically for the South and the fruit crops which grow best here.

Taking Care of Your Trees

Planting is only the beginning of the long and rewarding experience of growing your own fruit. Producing fruit is both an art and a science. We prune the tree or vine to look its best and to bear a full crop. This is a real art which you learn through study, practice, and time. And as you understand the science of *why* fruits grow, the art of growing them comes easily. (See the discussion of why fruit trees grow, beginning on page 15).

Producing fruit involves controlling weeds, killing insects, preventing diseases, applying water, fertilizing, pruning, and harvesting. We always want to harvest, but good fruit trees require culture, and a watchful eye.

I cannot overemphasize the importance of frequent visits to the orchard or tree.

> *If the eye is keen*
> *And labor long,*
> *Your harvests will be many,*
> *Your joy, great.*

Planning & Planting Your Family Orchard

A small family orchard can give you many hours of enjoyment and, if managed properly, a profit from sales. It may be one strawberry pot or a 10-acre orchard.

Before buying a fruit tree and sticking it in the middle of the back yard, sit down and plan the orchard. You have already made a wise move by buying this book and reading this far. Order several nursery catalogs from the list on page 87. Visit neighbors and see what is already growing well or, more important, not growing well in your community. Contact your county Extension agent and tell him your plans. He will become your most important source of horticultural information: He can supply you with a tremendous amount of information prepared for your state by specialists in all areas of fruit production.

THE PLAN

Once you have all the information before you, *make a plan*. Six months before you intend to plant the first tree is not too early. A checklist will help you grasp your specific situation (Figure 1). Draw a design showing where each tree of each variety is to be planted in the orchard. The smaller the orchard, the more important the design, because you want to use your space efficiently without crowding the trees.

The Orchard Size

If the orchard is to be partly commercial, it is critical that the size be within your economic and time constraints. If it is too large, which is very frequently the case, the chances for failure are greatly increased. It is far better to start any fruit planting small and increase the number of trees as you become familiar with the fruit's cultural requirements. More importantly, if the planting fails, great expense and disappointment can be avoided.

Location _Shongaloo, Louisiana_   Date _October 15, 1990_

Crop _Pecan_   Varieties _Choctaw, Sioux_
Crop _Peach_   Varieties _Springold Sentinel Harvester_
Crop _____   Varieties _Ranger Red Globe Loring_
Crop _____   Varieties _Dixiland, Redskin_
Crop _Pear_   Varieties _Orient, Ayers, Moonglow_
Crop _Apple_   Varieties _Starkrimson, Holland, Jerseymac_
Crop _Fig_   Varieties _Celeste_
Crop _Blackberry_   Varieties _Brazos, Comanche_
Crop _Plum_   Varieties _Morris, Methley, Ozark Premier_

Crop Total _7_   Variety Total _22_

Soil: Texture _Sandy clay loam_   Soil pH _6.0_
        Depth of Top Soil _6 inches_ Type Sub-Soil _red clay_
        Surface Drainage _good_   Internal Drainage _fair_

Climate: Record Minimum Temp. ____ _2°F_
          Winter Chilling Received ____ _800 hours_
          Average Last Spring Frost ____ _March 1_
          Average First Fall Frost _November 22_
          Number of Growing Days ____ _260_

Site: Frost Pocket ___ _none_ ___ Highway _blacktop_
        Animal Problems _deer, gofers_ Tax Problems _none_
        Disease History _none_   Available Water _city_

Orchard Plans:

        Design ____ _square_
        Spacing ____ _25 x 25 feet_
        Plant Source ____ _mail order nurseries_
        Tree Delivery Date ____ _January 15, 1990_
        Ground Preparation ____ _October 30, 1989_
        Irrigation ____ _drip irrigation_

Marketing Plans: Family Use ____ _20%_
                   Roadside Retail ____ _80%_
                   Wholesale ____ _none_

Figure 1. A typical checklist for planning the family orchard. (Also see page 93.)

Home orchards should also begin small and expand with the crops that produce the best. You may be surprised at the performance of a crop that you did not expect to do well. Also, you may find that what you want to grow simply will not do well.

The Orchard Design

There are several basic systems used for laying out orchards. They include the square system, the rectangle system, the diagonal system, and the triangle system (Figure 2). The square is by far the most commonly used, probably because people don't plan or consider other systems. The rectangle system is simply the square system with temporary or filler trees being used in one direction of the rows. As the trees crowd, the filler trees would be removed.

The diagonal system is also a modification of the square system. With it, permanent trees are planted on the square and temporary or filler trees are planted in the middle of the square. The diagonal system allows more effective use of your space.

The triangle system is very difficult to lay out and is very seldom used. The rectangle and diagonal systems will allow you to plant more trees in your space.

A planting plan for a ½-acre family orchard is illustrated in Figure 3. It includes the square, diagonal, and rectangle systems. Muscadines, blackberries, and strawberries are also presented in rows.

Plan Your Work Load

Keeping up with a small orchard can be lots of fun or very hard work. The difference depends on your labor force and the size of your orchard. Don't plant more

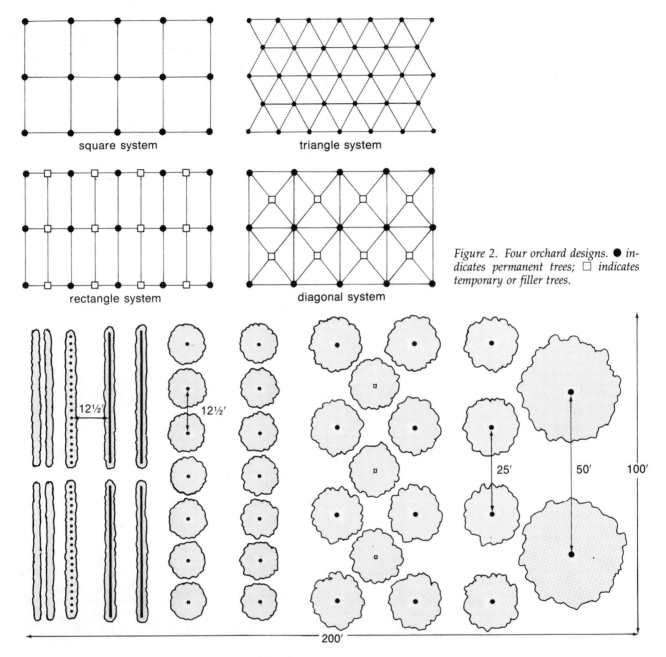

square system

triangle system

rectangle system

diagonal system

Figure 2. Four orchard designs. ● indicates permanent trees; □ indicates temporary or filler trees.

12½'   12½'   25'   50'   100'

200'

Figure 3. A typical ½-acre family orchard.

trees than you can maintain in excellent shape; you can always expand the orchard later, when you will know more about what is required and what you like to work with.

Evaluate Your Site

Few people select their home site with fruit culture specifically in mind. Most of us have to live with the site and climate we have. We can, however, select the crop and varieties. It is important to understand your location and its potential for growing fruit. Valleys and low places are serious frost pockets, but the soil there is usually very fertile and deep. Hillsides, on the other hand, have excellent air and surface water drainage, but usually have thinner soil. Check the chart on page x for the approximate number of growing days and the average first fall and last spring frost dates in your area. Determine the hours of chilling (total hours per winter that temperatures are below 45°F and above 32°F) received in your area before you select your crops and varieties (see Figure 10, page 13, for this information).

Evaluate the Soil

Poor soil drainage, both surface and internal, is the number one limiting factor in successful fruit culture. Evaluate the fertility and drainage of your site. Have a soil sample analyzed through your county Extension agent's office. To determine the internal drainage of your soil, dig an 8-inch hole 32 inches deep and fill it with 5 gallons of water. If the hole is empty in 24 hours, the soil has good internal drainage. If it requires 48 hours to drain, the internal drainage is poor but adequate. If the water doesn't drain, grow your trees in special raised beds made with the adjacent topsoil. It is also important to check the subsoil at your location. A red subsoil indicates good subsoil oxygen. A gray, yellow, or blue subsoil indicates no oxygen and can cause problems.

Design and Install a Drip Irrigation System

Drip irrigation or microsprinklers are important new developments in fruit culture. Growers in the South have had tremendous tree growth and production in very short periods of time with drip irrigation or mi-crosprinklers. Each system gives the trees their exact water requirements each day of the growing season.

The irrigation system should be installed before the trees are planted. The basic system includes a water source, filter, pressure regulator, pressure gauge, cut-off valve, main line, lateral line, and emitters or mi-crosprinklers (Figure 4). The drip system will only operate with very clean water. Silt, sand, algae, and insects must be filtered out to prevent the emitters from clogging. Microsprinklers can also clog, but it is less common. Microsprinklers distribute water over a wider area and require more water per minute to operate properly. Some growers begin with drip emitters and later convert to microsprinklers. If this is to be practiced, design the system to deliver the volume of water that will be needed by the microsprinklers and not the drip emitters.

The main line should be rigid PVC installed below ground. Laterals carry the water to each tree. The lateral lines may be installed either above or below the ground. When they are installed above ground, soft, special polyethylene pipe treated for resistance to damage from infrared rays is generally used.

When the laterals are installed below the ground, rigid PVC is a *must*, since gophers are such a problem in all

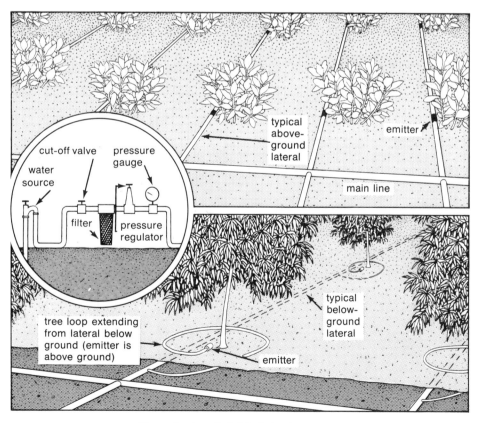

Figure 4. A typical drip irrigation system.

areas of the South. They'll easily tear holes in the soft polyethylene pipe. A tree loop feeds off of underground laterals (see Figure 4). The loop extends from the below-ground lateral up to the surface, so the emitter or microsprinkler rests above ground.

Night watering attracts rats and other rodents. They get accustomed to drinking from the emitters, and when the water is turned off, they'll destroy the emitters or lines in trying to get a drink.

When planning your drip system, remember that fruit trees should be irrigated according to their water needs and the drainage of the soil. A 1-year-old tree will usually require 1 gallon of water per day. If the soil is very sandy, the temperature and wind high, and the sunlight intense, the tree can use 2 gallons per day.

Take care to prevent excessive watering with drip irrigation, especially if yours is a heavy clay soil with poor internal drainage. Overwatering is a very common tree killer when an irrigation system is used. As many as four negative results can occur when oxygen is forced out of the soil air because of excess irrigation. These include young roots dying, water absorption stopping, salts becoming toxic, and soil carbon dioxide turning into toxic acid.

A 2-year-old tree usually requires 2 gallons of water a day; a 3-year-old tree will usually require 4 gallons of water daily. The minimum drip irrigation water requirement for mature bearing trees is presented in Table 1.

Table 1
Minimum Mature Plant
Irrigation Water Requirements

Crop	Gallons per Week
Pecan	350
Orange or Grapefruit	210
Avocado	210
Peach	105
Apple	105
Pear	105
Sour Cherry	105
Apricots	105
Plum	70
Fig	70
Satsuma, Kumquat, or Meyer Lemon	70
Persimmon	70
Muscadine	70
Cherry Plums	35
Blueberry	35
Grapes	35
Blackberry	14

These rates may be relatively low for arid situations, but for most of the South, where over 30 inches of rainfall usually occurs annually, these rates will be adequate to mature the crop and sustain good tree health.

Be sure your water source is ample enough to meet these minimum irrigation requirements.

PLANTING

Most fruit trees are planted in a similar manner, so the procedure can be described collectively.

First, and most important, never allow the roots to dry out, not even for 5 minutes. Take extreme care to insure moist (but not *soaked*) roots at all times.

Often, trees you buy at your local retail nursery are in plastic bags. When this is the case, purchase early in the winter, and plant or heel-in the trees immediately. Likewise, mail-order trees should either be planted immediately or heeled-in. Heeling-in is accomplished by covering the entire root system with moist soil. This does not mean covering the roots with straw or burlap. Also take the trees out of the shipping package before heeling-in. Figure 5 illustrates how to heel-in fruit trees between purchase and planting. Keep the heeling bed moist, but don't saturate it with water.

Stake your new orchard site according to your orchard design *before* the trees arrive for planting. If a small orchard is your intent, a steel tape and good eye are all you need. If you're planting an acre or more, you'll have to set a straight base line with stakes before the rows are laid off. All rows can then be laid off at right angles from the base line with a steel tape or chain. Check and double-check the rows in several directions to make certain they are straight: You won't be able to straighten them

Stake out your orchard design before you plant the trees. You want your rows to be straight, and this is the way to insure they will be.

after the trees are planted—rows can be with you a long time.

To plant the tree, dig the hole only slightly larger than the root system (Figure 6). Use a planting board to insure that the tree is planted at the exact spot the stake was placed. The planting board also helps to insure the

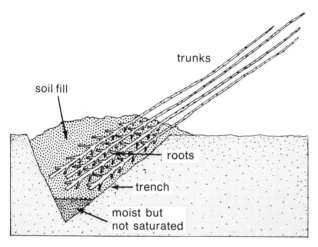

Figure 5. *Dormant trees heeled-in for temporary storage.*

tree is planted at exactly the same depth as it grew in the nursery row. The nursery soil line is usually rather easy to find because the color of the stem is different below the line, where the root zone begins.

Do not twist or turn the roots in the hole. Prune off all damaged or broken roots when planting. The tap root of tap root trees should sit on the bottom of the hole. With a hoe handle or foot, tightly pack the soil that came out of the hole back into the hole. Don't put commercial fertilizers in the hole when planting; they can cause salt burn. One cup or a handful of cottonseed meal or well-decayed barnyard manure can be placed in the hole. Do not use potting soil, sand, or topsoil to fill the hole as they can then become watersinks, which are saturated during rainy periods.

Build a small well around the tree to hold the irrigation water, and apply 5 gallons of water to the tree immediately after it is planted, even when a drip irrigation system is installed. Cut the stem back severely to keep the top in balance with the roots (most small roots will die in planting). All fruit trees will profit from heavy pruning at planting. If the top isn't cut back, too many buds will grow and transpiration will be greater than the amount of water the new roots can absorb.

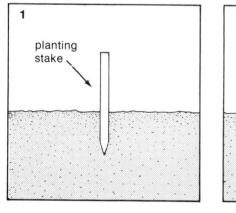
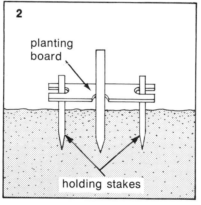
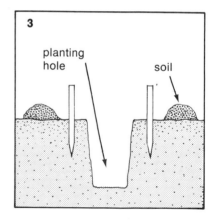
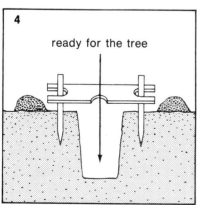
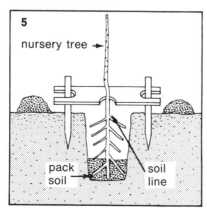
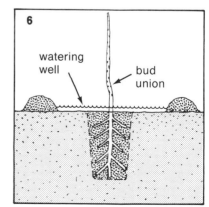

Figure 6. *Steps in using a planting board.*

If your trees are planted immediately, without drying out, and if the top is cut back severely, your trees will have strong growth the first year.

Southern Soils

Fruit trees will grow in almost every soil of the South. Some soils are excellent; others require special attention. The best soil provides the plant with good drainage, mineral nutrients, water, oxygen, microorganisms, and anchorage. A good soil is easy to till, does not hold water, and allows unrestricted root penetration.

Mineral Nutrients

The soil supplies plants with carbon, hydrogen, oxygen, nitrogen, phosphorus, potassium, calcium, sulfur, iron, manganese, magnesium, copper, zinc, boron, and molybdenum. The plants combine carbon, hydrogen, and oxygen to form sugar.

Nitrogen is the most important of all minerals and is necessary for the manufacture of protein and plant tissues. Phosphorus is necessary for good root development. Potassium is essential to the plant because it regulates the absorption of many elements and the movement of water throughout the plant.

Fertilizers

You'll need to fertilize your soil to see your fruit trees make optimal growth. You will find two types of fertilizers: complete and nitrogenous.

A *complete fertilizer* is composed of nitrogen, phosphorus, and potassium. It supplies these three minerals in one application. Formulations such as 10-10-10 are 10 percent nitrogen, 10 percent phosphorus, and 10 percent potassium (the rest is inert filler material). A 50-lb sack of 10-10-10 will contain 5 lb of nitrogen, 5 lb of phosphorus, and 5 lb of potassium.

A soil analysis processed through your county Extension agent's office includes a fertilizer recommendation. However, 1 lb of fertilizer per inch of mature tree trunk diameter is a good rule of thumb.

If a soil analysis shows that nitrogen is the only element needed, use a *nitrogen fertilizer*—it's less expensive and contains more elemental nitrogen.

To fertilize properly, apply a complete fertilizer in late winter before growth begins, and side dress with a nitrogen fertilizer later in the season.

Phosphorus fertilization can do more harm than good when the soil pH is above 7.8. The excess phosphorus ties up zinc and iron in the soil and in the plant. Do not apply phosphorus unless it is at a very low level in the soil. Continued use of complete fertilizers on high-pH soil can cause an accumulation of excess phosphorus. Phosphorus, unlike nitrogen, is not mobile in the soil and will not leach out.

When purchasing fertilizers, figure the value according to what the nitrogen costs per pound. Table 2 lists the major types of nitrogen fertilizers, with price comparisons.

Table 2
Cost of Nitrogen (per pound)
from Different Fertilizers

Type Fertilizer	%N	lb Nitrogen per 50-lb Sack	Cost per Sack	Cost of Nitrogen
10-10-10	10	5.0	$4	$.80/lb.
Sodium nitrate	16	8.0	$4	$.50/lb.
Ammonium sulfate	21	10.5	$4	$.38/lb.
Ammonium nitrate	33	16.5	$4	$.24/lb.

Soil Texture

Soils are composed of billions of tiny particles of three sizes: sand, silt, and clay (Figure 7). Large sand particles facilitate good drainage, but they cannot hold mineral elements well. The best soil is called a loam and will have clay or silt particles for mineral-holding and sand particles for good drainage.

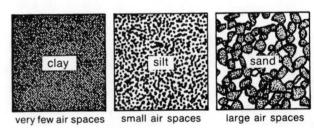

clay	silt	sand
very few air spaces	small air spaces	large air spaces

Figure 7. Soil particle sizes and texture.

Sandy Soil. The major commercial fruit industries of the South are located in areas with sandy soil. This is simply because sandy soils drain well and do not waterlog during rainy seasons. Sandy soils are also very porous, allowing a good oxygen supply and easy root penetration.

Clay Soil. There is very little air in clay soils, and they become waterlogged during wet seasons. During dry weather, clay soils are extremely hard. Each of these disadvantages demands special attention if fruits are to grow successfully in heavy clay soils.

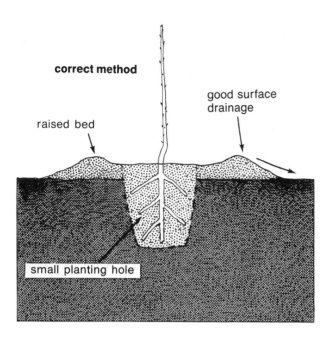

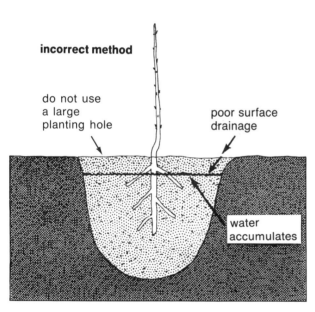

Figure 8. How to plant in clay soil.

The trees should be irrigated frequently with small amounts of water to prevent waterlogging. In flat areas, place the trees on raised beds to insure maximum surface drainage. Don't dig a very large hole for planting; this creates a water tank and will drown the tree. Figure 8 illustrates how to properly plant a fruit tree in clay soil on a flat location.

It is almost impossible to grow fruit trees on clay soils, which have a high sodium content. These soils should be avoided. Also, use as little salt water or city water as possible for irrigating fruit trees on clay soil.

Organic Matter

Organic matter helps the soil hold water and nutrients. Centuries of warm winter temperatures and long seasonal feeding by micro-organisms have greatly reduced the natural level of organic matter in most southern soils. The small amount present is very important for helping hold mineral nutrients. Typical southern sandy soils are usually less than 1 percent organic matter. This is quite different from the soils of the midwest, which have low microbial activity during cold winter months. Southern fruit growers need to fertilize to make up for the soils' low natural fertility.

The organic matter of southern soils can be only temporarily improved with annual applications of cottonseed meal, barnyard manure, cover crops, or other similar material. These products add little nitrogen to the soil, but they do add humus, which greatly improves the soil's nutrient-holding capacity. Continuous annual organic applications or mulching will be needed for sustained soil improvement.

Soil pH

Soil pH is a simple code from 1 to 13 used to describe the acidity or alkalinity of the soil, a pH of 1 being extremely acid and a pH of 13 extremely alkaline. Soils of the humid southeast are commonly acid; soils of the arid southwest are alkaline. Plants seldom live in a soil that has a pH below 4.0 or above 8.5. Most fruit crops grow best in soils with a pH ranging from 6.0 to 7.5.

Highly Acid Soils. Soils with a pH below 5.5 are highly acid and usually have serious calcium shortage problems. Fortunately, this problem can be corrected by adding lime (calcium carbonate) to the soil. The lime adds calcium and raises the pH. Table 3 gives the lime requirements for different southern soils. The exact lime requirement for any specific site can be obtained by having a soil analysis processed through your county Extension agent's office.

Table 3
Approximate Quantities of Lime Required to Raise Soil pH from 5.5 to 6.5

Soil Texture	Oz per Sq Ft	Tons per Acre
Sand	1.1	1½
Silt	1.8	2½
Clay	2.5	3½

Highly Alkaline Soils. Plants grown in soils with a pH above 7.8 (highly alkaline) usually have serious zinc and

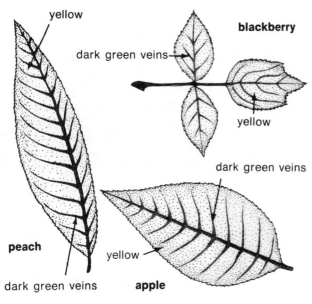

Figure 9. Iron chlorosis from alkaline soils.

iron deficiencies. Apples, peaches, and blackberries develop iron chlorosis (symptomized by very pale yellow leaves and dark green veins) from a shortage of iron (see Figure 9). Pecans will have zinc rosette and cannot make over 3 inches of shoot growth per year in highly alkaline soils because of an unavailability of zinc, which also causes pecan leaves to be very small (page 68).

Iron and zinc are usually present in alkaline soils, but due to the high soil pH they aren't dissolved in the soil solution. Unfortunately, there is little you can do to the soil to correct high pH problems. If you only have a few fruit trees, you can add organic matter or finely ground sulfur to the soil and lower the pH somewhat. However, this treatment is not nearly as effective as liming is for correcting highly acid soils.

Water

Most fruit crops require 30 to 50 inches of water annually. The trees need more water during early spring growth and later in the season as the fruit develops. Moderate amounts of water are required to maintain good foliage during hot southern summer months and some moisture is needed in the winter dormant months. No southern fruit crop can experience a 3-week drought without ill effects, and some form of irrigation will be needed. Drip irrigation (see page 5) or microsprinklers are excellent means of delivering water to fruit crops.

Water Stress Problems. Heavy rains for several days can kill trees if they are planted in clay soil in a flat, poorly drained location. Extended droughts, like those in the early 1950s, can kill fruit trees even in good soil. Fruit drop is a serious problem when fig, persimmon, and Navel orange trees become water-stressed. Drought followed by rain can cause serious fruit cracking problems with pecans, apples, peaches, and grapes. Extended winter droughts can seriously damage the trees' root systems and result in twig die-back or irregular bud break in the spring.

Salt Problems

Soils of the arid southwest frequently have salt problems. Concentrations of 3,000 ppm soluble salt will make fruit culture extremely difficult. Again, a soil analysis processed through your county Extension agent can tell you the salt concentration in your soil.

Only the more salt-tolerant crops such as figs, Vinifera grapes, pomegranates, and pecans should be attempted in soils with high salt concentrations. If your soil salt concentration is high, irrigate frequently to help reduce the build-up of salt following evaporation. This must be followed with occasional heavy irrigations to leach the accumulated salts from the soil.

If the soil sodium content is 250 ppm or more, internal drainage problems will occur. This can be corrected somewhat with gypsum (calcium sulfate). Soak 2 oz of gypsum into every square foot of the soil.

Never allow sprinklers to spray the foliage when the salt concentration is above 1,000 ppm. The photo (on page 70) shows salt burn on pecan foliage from sprinkler watering. The water and salts are carried into the leaves; when the water transpires, the salts remain and burn the leaves. Most city water contains high levels of sodium and chloride to kill bacteria. Contact your city water works office to determine the salt level of your water.

The Southern Climate: What it Means to the Fruit Grower

The weather determines where most fruit crops can be grown. Citrus is grown in Florida, Arizona, and Texas because of the infrequency of severe freezes. Apples are grown in the North because of good winter chilling and cool fall ripening temperatures. Vinifera grapes are grown in California because of the dry, mild climate.

The kind and quality of fruit we can grow in the South is influenced by temperature, moisture, wind, and hail.

Temperature

Fruits grow best at 70° to 90°F. Photosynthesis and optimum food production by the foliage occurs in the morning hours, before the daytime temperature moves above 90°F. Consideration is now being given to the idea of running fruit rows east and west, rather than north and south, in order to maximize light penetration down the rows while the growing temperature is optimum. Temperatures above or below this can inhibit normal growth and development of the plant. The orchardist usually considers lower temperatures as one of his most critical limiting factors. Winter temperatures below 0°F frequently kill mature trees. Fortunately, this seldom occurs in the South. However, early fall freezes commonly destroy fruit trees before growth slows down and the wood hardens. Still more important, late spring freezes destroy billions of fruit blooms in all areas of the South each year.

To be a successful fruit grower, you must manage your trees to reduce these cold-associated problems.

As a fruit tree grows through the spring, summer, fall, and winter, different growth processes are taking place. You can exert some control over these processes through site selection, variety selection, and cultural practices.

Spring Is the Beginning

Springtime is blossom time in the South. Fruits of all types respond to the warming temperatures and rains with a beautiful show of blooms. Unfortunately, late freezes sometimes follow.

Fruit trees respond to the warmth of spring with a beautiful show of blooms.

Warm temperatures and increasingly long periods of daylight stimulate the accumulation of growth-promoting hormones within the buds of the plant. Once enough warm hours have accumulated, the tree will bloom and grow. Ideally, this should occur relatively late or after the average frost date for your location. In the South, high numbers of warm hours can accumulate in January and February, setting the trees up for potential frost damage. Even in March, when many southern fruits bloom, several freezes may still occur.

Once the trees have bloomed and growth has begun, begin your cultural practices. Control weeds, which compete for soil minerals and moisture. Begin watering 1 week after growth begins. Fertilizers are needed to stimulate growth. The best growth period for all fruits is in the spring: Temperatures are in the optimum range, rain is abundant, insects and diseases are usually not much of a threat, and a good supply of stored food moves up from the roots.

The Long, Hot Summer

Once past the late freezes, fruit growth progresses normally until extremely high summer temperatures are reached in June or July. If no irrigation is available, droughts also reduce or stop growth. In general, growth is optimum and is seldom affected at temperatures below 95°F. Temperatures above this, combined with wind and high light intensity, can greatly increase the plants' water requirements. Growth will stop, and wilting indicates that the plant is not receiving sufficient soil moisture. Unfortunately, many fruit crops do not wilt to indicate a water shortage. If water is not supplied, desiccation, and in extreme cases, death, can occur.

Summer cultural practices will maintain healthy foliage for food manufacture and fruit development. Weeds can be very competitive and must be controlled. Insects are plentiful and have to be killed. Unnecessary shoot growth should be thinned out.

The most important summer cultural practice is water management. Use drip irrigation or microsprinklers to stimulate optimum leaf growth and fruit development during our long hot summer (see page 5).

Fall Ripening and Harvests

Cool night temperatures in the fall are very important for optimum fruit ripening. With cool fall temperatures, apples, pears, and citrus all ripen with beautiful color and proper sugar and acid levels. Southern apples do not have strong red color because the late July and early August nights are too warm. The Navel orange and Sat-

All deciduous fruit crops require a certain amount of chilling every year before they can bloom and fruit properly.

suma will develop higher quality and brighter orange color in the New Orleans and Mobile area than they will in the lower Rio Grande Valley.

Fall is the season for preparing the trees for winter freezes. If fruit trees are exposed to freezing temperatures before growth slows down, major freeze damage can result. Begin cultural practices that reduce growth 60 days prior to your first fall freeze date (see chart on page x). Irrigations should be very few so as to inhibit new growth. The foliage should be protected from disease and insects for maximum food manufacturing, fruit ripening, and cold hardiness. Excellent fall foliage is a must for optimum bloom and growth the next spring. Plants that are weak going into the winter are very susceptible to cold damage.

Winter Rest Period

Deciduous fruit trees go through a rest period in the winter. Temperature, day length, and hormones within the buds regulate the time deciduous trees lose their leaves in the fall. These hormones also keep the trees dormant in the winter and stimulate growth in the spring.

Apples, cherries, pears, plums, and most northern temperate fruit crops have an obligatory winter rest period. If these trees are *not* exposed to sufficient cold for a sufficient period, they will not bloom and grow normally in the spring. This is no problem in the North, but as we attempt to grow these fruit crops in the South, winter chilling becomes an important limiting factor.

All deciduous fruit crops and varieties have a specific chilling requirement. The chilling requirement for a variety is defined as the accumulation of hours of winter chilling below 45°F and above 32°F required by flower and vegetative buds for normal bloom and shoot growth in the spring. Northern varieties of pears, apples, peaches, cherries, and plums planted in the lower South will not survive. This is simply because they do not receive enough cold to satisfy the trees' chilling requirements. During mild winters these trees will not bloom normally and the tree will remain dormant until early summer, when a small shoot or several fruit will form on the very end of a long naked shoot.

Fruit trees differ in their chilling requirements, and so do the varieties within a species. This is indicated in Table 4. Fortunately, the USDA and state Agricultural Experiment Stations across the South have been breeding and selecting low-chilling varieties which now offer several previously unavailable varieties. Tremendous progress has been made in the introduction of new apple and peach varieties.

Figure 10 shows the approximate minimum hours of chilling received in the South. If a variety receives too much chilling for a specific location, it could bloom too early. If a variety doesn't receive enough chilling, it will have delayed growth in the spring. Ideally, a variety's chilling requirement should match the chilling hours the location receives. But a difference of 200 hours is common, with both problems and advantages resulting.

Minimum Winter Chilling Zones in the Southwest-Southeast

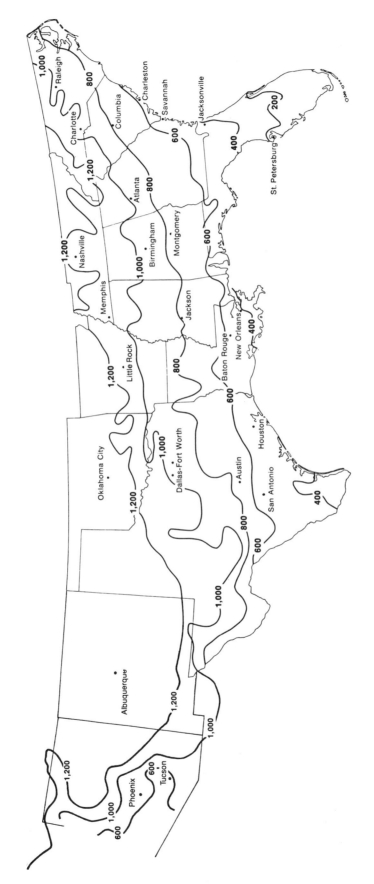

Figure 10. Approximate minimum hours of chilling (total hours per winter that temperatures are below 45° and above 32°F) received in various areas of the southwest-southeast regions of the United States.

Table 4
Common Chilling Requirements of Southern Fruits

Crop	Varieties	Chilling Requirements*
Peaches	Red Haven	950 hours
	Red Globe	800 hours
	Sentinel	750 hours
	June Gold	650 hours
	Rio Grande	450 hours
	EarliGrand	250 hours
Cherry	Montmorency	1,000 hours
Plums	Ozark Primer	750 hours
	Methley	600 hours
Apples	Golden Delicious	850 hours
	Starkrimson	700 hours
Pears	Maxine	850 hours
	Orient	650 hours
Pecans	Stuart	500 hours
	Desirable	400 hours
Grapes	American varieties	100 hours
	French varieties	None
Citrus	most varieties	None

* Accumulation of hours per winter that temperatures are below 45°F.

Winter Sun Scald. Freeze injury can occur near the ground line on the south or southwest side of southern fruit tree trunks (Figure 11). Direct sunlight and the absence of foliage and shade on this area of the tree retard the onset of the winter rest period. The lower trunk is heated in the daytime and frozen at night. This is especially common on young, non-bearing trees which are growing late into the fall and have very little old bark for protection. The problem can be reduced by painting the trunk with a whitewash or latex paint. One-year-old trees can be wrapped with a white plastic tree guard or aluminum foil to reduce winter sun scald injury.

Deep Winter Freezes. Southern fruit trees that are fully dormant and in the middle of the winter rest period can be injured, and occasionally killed, if the temperature drops below 0°F.

Moisture

Most fruit crops will grow and produce satisfactory crops with 30 to 50 inches of rainfall annually. One inch of water per week is the minimum and two inches is the optimum. With 32 weeks of growing conditions, 32 inches of water is needed and 64 inches is the optimum if the soil drains properly. Certain periods of the production season require more moisture than others. These are just prior to bud break, midsummer, and during fruit sizing. As the trees enter the fall, water should be reduced to help bring on the onset of the winter rest period. During the winter some soil moisture is required to prevent root desiccation.

Negative Effects of Rainfall. "When it rains it pours." The saying was probably coined by a fruit grower. Many phases of fruit production are conducted on a definite time schedule, and spraying, harvests, weed control, and pruning can all be delayed by rain. Areas along the Gulf of Mexico which receive over 60 inches of rain experience flooding and water drainage problems. Trees in these areas have to be planted on small ridges (Figure 8).

Rain can play havoc with harvesting. The home or commercial fruit grower should harvest the fruit only when firm ripe. Consequently, it must be picked within a very short period of time. Strawberries, peaches, and grapes should be picked within 24 hours of the optimum ripe stage. Rain at this time reduces fruit quality and fosters disease problems. Brown rot of peaches, black rot of grapes, gray mold of strawberries, scab of pecans, and melanose of citrus are all brought on by excessive rainfall. These diseases must be controlled with preventive fungicide sprays or the crop will be lost.

Wind and Hail

Continuous winds can make tree training extremely difficult. In some areas of Texas and Oklahoma winds can cause trees to lean at a severe angle. Winds can lay down apple and peach trees established under drip irrigation. Place drip emitters downwind from the trees and plant the rows in the direction of the prevailing wind.

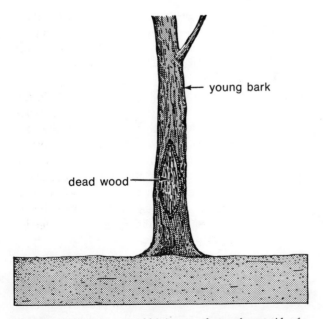

Figure 11. Winter sun scald injury on the southwest side of a fruit tree.

young bark

dead wood

There is little else you can do to correct the problem; fortunately, it is only serious on rare occasions.

Vineyard trellises can be blown over in Texas if intermediate posts are not used every ten vines. Texas winds also require each vine to be individually staked.

High winds frequently break limbs out of rapid-growing pecan trees. This is especially true of varieties that are difficult to train, such as Desirable and Wichita. It is seldom a problem on mature slow-growing trees or trees that have been trained properly to a modified central leader system. When grafting, always place the graft stick on the windward side to keep it from blowing out. Chemical sprays should be applied so that the wind blows the spray away from the applicator.

Sand is a problem associated with wind. Windblown sand can seriously damage peaches, nectarines, apples, and table grapes.

Some areas of the South receive considerable hail. If this is the case in your area, think twice—hail is very destructive to fruit trees.

Why Fruits Grow

All plants receive their energy from the sun. Special green structures within the leaves trap the sunlight, and in a fraction of a second the plant uses this light energy to transform water and carbon dioxide into sugar and oxygen. This sugar then becomes a basic building block for the entire plant and its fruit. It is a small wonder, then, that the juices of many fruits are extremely high in sugar (Table 5).

Table 5
Sugar Produced in the Juice of
Several Southern Fruits

Fruit Crop	Percent Sugar in Juice
Grapes	22%
Strawberries	18%
Peaches	10%
Apples	17%
Muscadines	20%
Navel Oranges	10%
Grapefruit	8%
Blueberries	15%
Blackberries	8%

To grow fruit successfully, you must understand the basics of plant growth. As Dr. Julian C. Miller once said, "We cannot truly learn *how* to culture fruit until we understand *why* fruit grow."

A fruit tree begins as a seed. The seed germinates and grows to form a seedling. The initial food for growth comes from carbohydrates stored in the seed. Once this food is exhausted, the small seedling absorbs sunlight as energy and begins to manufacture sugar on its own. The sugar is then combined with minerals absorbed from the roots and hormones made in the shoot and root tips to develop new leaves, roots, and stems. With seasons and time, the small tree will mature from a non-bearing juvenile to a fruit-bearing adult. To grow from a juvenile to an adult can require 18 months for strawberries and as long as 30 years for pecans.

Trees produce fruit to propagate themselves. Man interrupts this system and harvests the fruit for his personal use. In centuries of doing so, we have selected varieties and developed fruit production systems that allow us to grow fruit of the very highest quality in the shortest period of time.

The fruit tree, as well as vines, bushes, and other plants, is made up of four basic structures: roots, stems, leaves, and fruit (Figure 12).

Roots

Roots absorb water and mineral elements from the soil when air and oxygen are available in the soil. They anchor the tree in place, preventing it from falling over with heavy crops or winds. The roots also serve as a massive storage compartment for reserve carbohydrates manufactured each year by the leaves.

Root Systems. Pecans, persimmons, and apples have a large *tap root* which serves for food storage, anchorage of the tree, and deep water absorption. Peaches, figs, grapes, and citrus have a *fibrous root* system which thoroughly penetrates the topsoil. Strawberries, muscadines, and blackberries have very *shallow fibrous roots* which are extremely close to the soil surface.

Feeder Roots. Feeder roots develop extensively in the topsoil, especially close to the soil surface. The feeder roots are very important to the plant because they absorb most of the minerals. They demand special attention because of their importance and their closeness to the soil surface. Any damage to them from cultivation, flooding, drought, compaction, or freezing can be detrimental to the tree and its capacity to absorb minerals and produce fruit. Always remember that the feeder roots are just below the surface.

Root Hairs. These specialized cells on the feeder roots are the agents of mineral and water absorption. The absorption process is an actual growth process requiring oxygen immediately adjacent to the root hairs. Before mineral elements enter the root hairs, they must be dis-

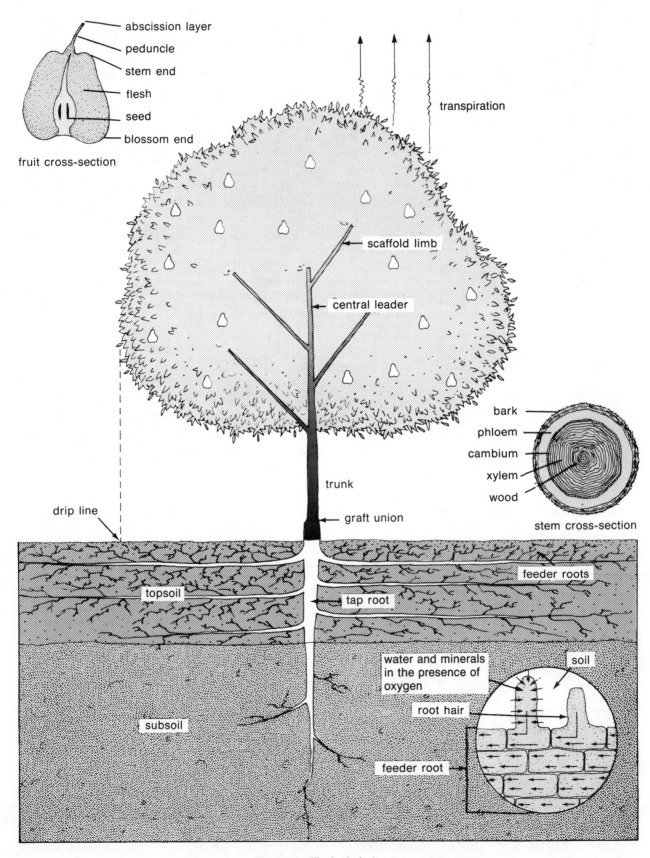

abscission layer
peduncle
stem end
flesh
seed
blossom end

fruit cross-section

transpiration

scaffold limb

central leader

bark
phloem
cambium
xylem
wood

stem cross-section

trunk

graft union

drip line

topsoil

tap root

feeder roots

subsoil

water and minerals in the presence of oxygen

soil

root hair

feeder root

Figure 12. The basic fruit tree.

solved in the soil water. A special ion-transfer system pulls the water and minerals into the root hair cells. Only a very small portion of the water remains in the plant. Most of the water moves through the tree and into the atmosphere by normal transpiration. A special pressure deficit created as water leaving the leaves through transpiration causes the water and minerals to be pulled up to the leaves through the wood or xylem tissue. If the temperature, light intensity, and wind are high, transpiration can be very great. A mature, bearing pecan tree can easily transpire over 100 gallons of water daily on a hot, clear, windy, southern day! So, you can see the importance of roots and their job of absorbing large volumes of water from the soil in the presence of soil oxygen.

Stems

Stems transport the water, minerals, and oxygen up to the leaves. Once these components are manufactured into food in the leaves, the stems again transport the food throughout the plant. The water and minerals move upward through the inside of the stem in special *xylem vessels*. The food and hormones manufactured in the leaves are moved throughout the plant through the outer area of the stem in *phloem tubes*. A special layer of cells, called the *cambium*, separates the xylem from the phloem. As the xylem ages it becomes wood; as the phloem ages it is pushed outward as bark. Each year, new stem growth originates from the cambium, with wood forming to the inside as annual rings, and bark to the outside.

Leaves

Leaves are food factories. By absorbing sunlight, sugar is manufactured through the process of photosynthesis in the leaves. Leaves must remain healthy throughout the growing season if the tree is to manufacture enough food. If insects, diseases, heat, drought, cold, or chemicals damage the leaves, the crop and occasionally the plant can be lost. Good leaf health is a very important part of successful fruit culture. All good fruit producers strive to maintain healthy leaves throughout the growing season and up to the first frost. When they do so, their trees produce excellent crops of potentially high-quality fruit *every* year.

Fruit

A tree grown from seed is called a seedling, and it develops through a juvenile stage before it bears fruit.

Once the seedling bears fruit, the wood on which the fruit is born is called adult. If enough stored food is available, the adult wood will continue to bear fruit. Nursery trees of a specific variety are always grafted from buds collected from adult wood. This means a nursery-budded tree will not have to grow through a juvenile phase.

A tree planted from seed will have to grow through a juvenile phase and will require much longer to come into production. Figure 13 illustrates the adult and juvenile zones of two pecan trees, one grown from seed and one a grafted tree. Shoots in the juvenile zone frequently have characteristics which are different from the typical adult. Juvenile citrus seedlings will generally have more thorns than grafted adult trees. Once a shoot on a seedling tree bears fruit, all growth from that point upward is adult.

Juvenile trees develop into central leaders, with only one dominant trunk, without any training because of a natural hormonal control. However, adult nursery-budded trees require very specific central leader training, which is difficult, expensive, and time consuming.

Extremely fast-growing trees, either adult or juvenile, will not bear fruit because the trees' manufactured food is used in new growth and does not accumulate in storage areas of the roots and stems. With time, the number of shoots on a tree will increase, growth will decrease, stored food will accumulate, and production will begin. The length of time normally required for adult grafted nursery trees to bear is presented in Table 6.

Table 6
Time Required for Grafted Fruit Trees to Bear

Fruit Crop	Approx. Time to Bear
Strawberries (fall planted)	6 months
Strawberries (refrigerated)	7 months
Strawberries (winter planted)	14 months
Figs	2 years
Blackberries	2 years
Plums	2 "
Grapes	3 "
Peaches	3 "
Satsumas	3 "
Nectarines	3 "
Muscadines	4 "
Blueberries	4 "
Persimmons	4 "
Apples	4 "
Oranges	4 "
Pears	5 "
Sour Cherries	5 "
Apricots	5 "
Grapefruit	5 "
Pecans	7 "

Grafting Fruit Trees

All seedling trees are different and will bear fruit different from the mother trees' fruit. Some trees grown from seed without grafting have produced relatively fair-quality fruit; however, 99 percent of all seedling trees have inferior fruit and require a long time to bear.

As an example, Professor P. L. Hawthorne of Louisiana State University at Baton Rouge, Louisiana, has bred and evaluated over 15,000 peach seedlings and has found less than 10 worthy of being named as a variety to propagate commercially. For this reason, purchase grafted trees.

A system which is gaining in popularity, especially for small pecan orchard growers, is to plant a juvenile seedling tree and, after two to five years, graft the adult wood of a desired variety onto it.

All grafted trees of a specific variety are genetically identical. The rootstocks of these grafted pecan trees are 3 years old. The tops are 1 year old.

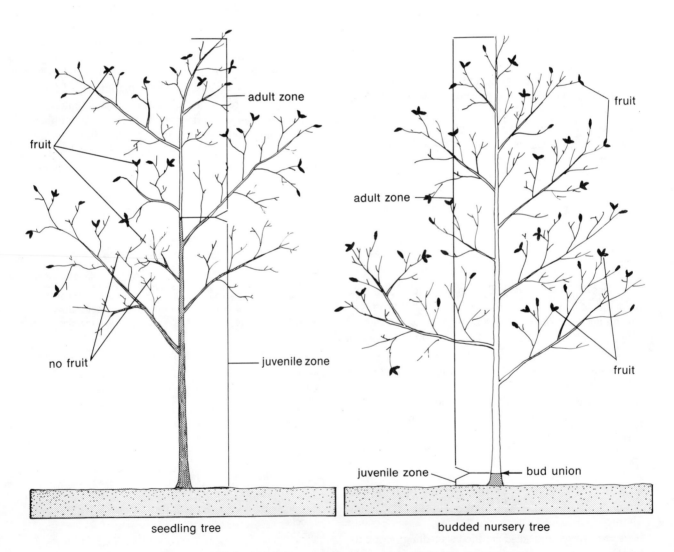

Figure 13. Adult and juvenile zones of seeding and budded nursery pecan trees.

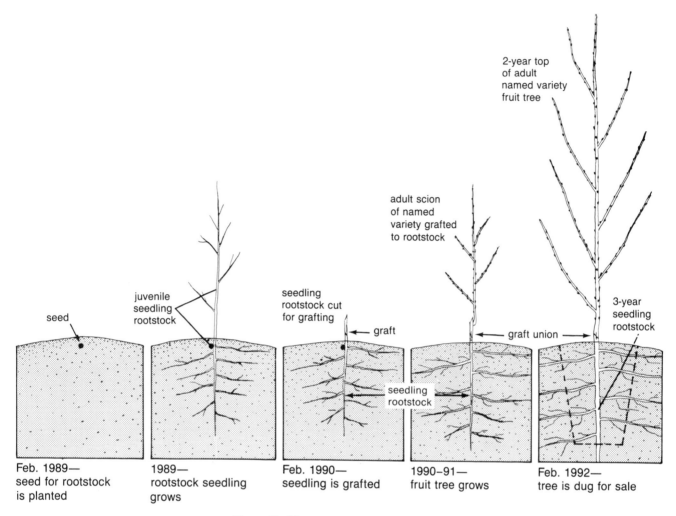

Figure 14. The way nursery trees are grown.

All grafted trees of a specific variety are genetically identical. Every Choctaw pecan tree in the world can be traced back to the original tree bred and selected by L. D. Romberg at the USDA Pecan Field Station at Brownwood, Texas. When a Choctaw bud or graft is grafted onto a seedling tree, the limbs that grow from the graft will be adult Choctaw limbs.

Nursery trees are grafted close to the ground in rows when the rootstock seedlings are only 1 or 2 years old. Figure 14 illustrates a typical nursery grafting procedure.

Nurserymen and horticulturists identify all grafted trees of one variety as a clone. Trees, vines, and bushes grown from a stem or root cutting also constitute clones. Each variety of these plants will have identical genetic material. Several important fruit varieties, such as Thompson Seedless grape and Navel orange, do not have true seeds and must be propagated by stem cuttings or grafting.

Fruit Seeds

The seeds produce hormones which stimulate an increase in the size of the fruit and prevent premature fruit drop. Fruit varieties with parthenocarpic fruit (fruit that develops without sexual fertilization of the mother) do not have true seeds and require special care to prevent premature fruit drop from the tree. Common figs, Navel oranges, and some Japanese persimmons are parthenocarpic and have major fruit drop problems.

Fruit Quality and Harvest

Leave the fruit on the tree until firm ripe. This is especially necessary for those crops which are *non-climacteric* (fruits that will not ripen or improve in quality after they are picked from the tree). *Climacteric* fruit crops can be picked when the fruit reaches full size or horticultural maturity. These fruit will continue to ripen after they are picked.

Table 7 lists the various crops and how they ripen. During the ripening process, starch will convert to sugar, thus greatly improving the dessert quality of the fruit. The ripening process is initiated by ethylene gas produced inside the fruit. Ripening will continue after harvest for climacteric fruit, but not for non-climacteric fruit.

Table 7
How Various Fruit Crops Ripen

Ripens off or on the Tree (Climacteric)	Ripens on the Tree Only (Non-Climacteric)
Avocados	Peaches
Apples	Strawberries
Pears	Blackberries
Plums	Grapes
Persimmons	Muscadines
	Figs

Pest Control

Never spray chemical pesticides unless you have to; if you do, check with your county Extension agent to be sure you use the right spray.

Many pests attack fruits in the South, and each must be dealt with lest you lose your crop. If totally disregarded, many pests can kill a tree or wipe out your orchard. Close observation is the key to effective pest control. Check your trees daily.

NATURAL PREVENTION

Many fruit pests can be controlled through natural means.

Variety resistance is by far the best means of avoiding many serious pests. Our hats are off to plant breeders who have selected varieties for the South which are naturally resistant to many pests here. The recent popularity and strong potential for muscadines and rabbiteye blueberries is due in large part to their natural resistance to insects and diseases.

Sanitation is extremely important. Pay constant attention to your trees and fruit, removing sources of disease,

infection, and insect build-up. The orchard should be as clean as possible. Ripe fruit should be picked and never allowed to fall and accumulate on the ground. Don't allow weeds to grow under the trees. Prune out diseased or dead wood and remove diseased, mummified, or dried-up fruit from your trees whenever you see it. Large insects can be picked from the tree and killed if a large population does not exist.

For hundreds of years sanitation was the only means of combatting pests. Its effectiveness is still important today. Never spray unless it is absolutely necessary.

CHEMICAL CONTROL

Agricultural chemicals have advanced southern fruit culture as much as any single cultural practice in the last 20 years. Today's fruit grower can depend on reliable chemicals to kill insects, prevent diseases, and control weeds. First, however, you must know the basic types of chemicals and what they can and cannot accomplish.

Insecticides

Insecticides are chemicals which kill insects and similar pests such as spiders and mites. They kill by three different methods: *stomach insecticides* kill when the pest eats foliage or fruit that has been sprayed; *contact insecticides* simply kill the insect on contact; and *systemic insecticides* are absorbed by the plant, translocated to all its parts, and kill the insect when it feeds. Many insecticides kill by a combination of these methods.

Oil sprays kill scale insects when used in the dormant season. Special summer oils are available for controlling scale, mites, and white flies on citrus trees during the growing season.

Chlorinated hydrocarbons such as methoxychlor are available for use on fruit trees. This is an old type of insecticide. There are very few federally approved chlorinated hydrocarbons available for use by fruit growers.

Organic phosphates such as malathion and diazinon are very common fruit insecticides. *Carbamates* are represented by sevin which, though very safe, will not control aphids, a major fruit pest. *Sulfur* is used to a lesser degree to control mites.

Chemicals from plants such as *pyrethrum*, *rotenone*, and *nicotine* are also commonly used to kill insects feeding on fruit trees. And a bacterium, *Bacillus thuringiensis*, is used to kill certain caterpillar insects.

Fungicides

Fungicides are chemicals that *prevent* fungus diseases. They do not kill or destroy the fungus and should be applied at the first sign of fungus development if they are to be effective. Serious fungus diseases such as grape black rot and peach brown rot can be expected under humid conditions; you'll need to spray frequently to prevent them.

Bordeaux mixture is a combination of copper sulfate, lime, and water, and has been used as a fruit fungicide since the late 1800s, but newer chemicals are more effective and much less damaging to the plant. *Carbamate* fungicides such as ferbarn, captan, zineb, and others are effective fruit fungicides. *Benomyl* is a relatively new fruit fungicide that is extremely effective against many major fruit fungus diseases.

Herbicides

Hand cultivation and mulching are good methods for controlling weeds. In many instances, however, they are ineffective and chemicals are required.

Post-emergence herbicides such as glysophate, dalopon, paraquat, poast, fusilade, and 2,4-D kill weeds through contact action. Spray the weeds until wet while they are growing rapidly. Never allow the herbicides to come in contact with the foliage or bark of the fruit trees. If post-emergence herbicides are misused, they'll kill the fruit trees. Poast and fusilade are unique in that they only kill grasses and do not harm broadleaf species. This is a new and functional tool in fruit culture. 2,4-D is the opposite in that it kills broadleaf species and not grasses; consequently, it should not be used near fruit trees.

Pre-emergence herbicides such as simazine, treflan, surflan, and diurone prevent weed seeds from germinating. These chemicals are effective only if no perennial grasses such as Johnson grass, Bermuda grass, or nut grass are present. The area to be treated must be perfectly clean and weed free when the chemical is applied. They can be applied in the fall or spring and can prevent weed seed germination for 2 to 3 months. Some pre-emergence herbicides can be absorbed by the fruit tree and result in tree damage if the proper rate is not used.

Other Chemicals

Antibiotics, soil fumigants, growth regulators, and other chemicals are federally approved for use on fruit trees. However, most of these are of minor value compared to insecticides, fungicides, and herbicides. Rooting hormones such as indole butyric acid (IBA) are very important in rooting hardwood and softwood stem cuttings. Gibberellic acid can be used to increase fruit size on table grapes.

Laws Regulating Chemical Use

Federal and state laws currently regulate the use of agricultural chemicals.

General classification pesticides are safe chemicals and can be purchased and used by persons without the purchaser or applicator being a certified pesticide applicator. Some important fruit chemicals can only be purchased and used by a certified applicator.

Restricted use pesticides can be purchased and used by certified private applicators only. Obtaining certification is not difficult but you'll be required to receive training on the correct use of restricted use pesticides (see your county Extension agent for details). All such chemicals bear this sign:

RESTRICTED USE PESTICIDE
For Retail Sale To And Application Only
By Certified Applicators Or Persons
Under Their Direct Supervision

The Pesticide Label

You must follow certain precautions when spraying your fruit trees. Always read and follow the instructions on the pesticide label. They will tell you the chemical's name, the active ingredient, directions for use, pests controlled, crops to be used on, hazards, toxicity signs, and precautions. The product label is the *best* source of information on use of the chemical.

Always check the toxicity of the chemical by one of the signal signs:

DANGER   = Highly Toxic
WARNING = Moderately Toxic
CAUTION = Low Order of Toxicity

In addition, *always make certain the chemical to be used includes the crop and pest on the label.*

Sprayers

Various equipment is available for your spraying requirements. The most expensive equipment is not necessarily the best for your job. The thing to remember is that insecticides and fungicides are most effective when you cover all of the fruit and foliage.

A *compressed air sprayer* can be used effectively on all crops except large trees such as pecans, citrus, avocados, apples, and pears. Exact chemical concentrations can be mixed and the spray is easily directed onto the plant from a distance of up to 6 feet. You should only purchase a polyethylene or stainless steel compressed air type sprayer. The galvanized type always has stopped nozzle problems once the tank begins to rust, and it *will* rust. The major disadvantage with compressed air sprayers is that they have a limited tank capacity and spray distance.

Trombone or slide sprayers can reach slightly further than the compressed air sprayers but are more difficult to use, as you have to carry the chemicals in a separate container.

Hose-on sprayers are excellent for spraying small fruit orchards. They are inexpensive, easy to operate, and spray a fair distance. A disadvantage is that they do not make exact chemical concentrations and the suction is occasionally lost. These sprayers are excellent if you have several large trees.

Small power sprayers are available which can deliver as much as 5 gallons per minute at 200 lb-per-square-inch pressure. Most of these sprayers have a 25-to-50-gallon tank and small wheels for easy movement in the yard or orchard. They are expensive, ranging from $400 to $900 each.

Skid sprayers are commonly used by commercial fruit producers before their trees begin fruit production. These sprayers are also excellent for the home fruit orchard. They can be operated out of a pick-up truck pulled by a riding lawn mower, or filled in-place near the trees to be sprayed. Tank sizes range from 50 to 200 gallons. The pump can deliver relatively high pressure and volume. Prices range from $500 to $1,000 each.

Larger sprayers can be evaluated by the pressure and gallons of water per minute delivered. Typically there are three ranges: 200 psi and 5 gallons per minute; 400 psi and 10 gallons per minute; and over 500 psi and over 12 gallons per minute. All three of these pump capacities can be attached to plastic, fiberglass, or stainless steel tanks varying in size from 100 to 500 gallons. The price ranges from $2,500 to $7,000 each. If possible, it is best to have a mechanical agitator to keep the spray solution well mixed. However, a high-volume solution bypass agitator can be functional. You should never spray any chemical without good agitation. Oil and herbicide sprays can kill your trees if proper agitation is not maintained.

Never purchase a steel tank sprayer, because it will have continuous rust, clogging, and calibration problems.

How To Spray

Spraying is a simple, safe, and necessary part of southern fruit culture. Spray *only when needed* to kill insects, prevent diseases, or control weeds. Wear plastic gloves and be certain your sprayer is operating properly before mixing your chemicals. Pour out the chemical very carefully. Mix the chemical immediately before spraying and not earlier; otherwise strength and effectiveness can be reduced significantly. Do not use strengths stronger than specified on the product label—increasing the concentration can increase the toxicity of the spray and also damage the trees.

Spray when wind drift is lowest. Apply enough spray to cover the foliage and see a little run-off from the leaves. Do not soak the tree. Early morning, late afternoon, or evening is the best time to spray.

Children and Spraying

Youngsters are fascinated by spraying. Be sure they understand the seriousness and danger of poisonous pesticides. Keep them away while mixing and spraying. Store the chemicals in a locked compartment out of their reach. Never purchase more chemicals than you can use in one season. The chemicals' effectiveness will deteriorate during the off-season and disposal can be a problem.

If the label directions are followed, agricultural chemicals are an important aid to successfully growing top-quality fruit. In many instances, their use is essential if any crop is expected.

Safety Precautions

Chemicals can be extremely dangerous before they are mixed in water to become a spray.

1. Always use rubber gloves and at least a paper nose/mouth respirator when handling the pure chemical.
2. Always wash your hands immediately after mixing the chemical in the water.
3. Always avoid smoking or eating during or immediately after mixing or spraying chemicals.
4. Never stick your head or face into a sprayer to see if it is clean or empty.
5. Change clothes and bathe immediately after spraying and never wear clothes which have not been washed after spraying.

Stone Fruits: Peaches, Plums, Apricots & Sour Cherries

PEACHES
Prunus persica

Peaches have come to the South from Persia and China, and since colonial days southerners have also grown a small seedling clingstone peach, called the Indian peach, which reportedly came from Spain.

Peaches have always been a major commercial crop in the South. In the early 1900s over 100,000 acres of Elberta peach trees were planted here. The Elberta was later replaced by some 50 outstanding varieties which ripen from the first week in May until September. South Carolina and Georgia are important commercial peach states, and considerable potential exists for other southern states, especially for small, family-operated peach orchards of ten acres or less.

Peaches are an excellent fruit for the orchard, but if you're only going to plant one tree, select another crop such as plums or pears because peaches require very close attention and are unattractive if planted alone.

Climate

Each peach variety requires a minimum number of hours of winter chilling below 45°F before it will grow normally. The South receives varying amounts of winter chilling (as indicated in Figure 10, page 13), so select varieties whose chilling requirements match the hours of chilling in your area.

High humidity fosters disease and insect problems, and frequent sprays are required to overcome them.

Peaches tolerate the coldest southern temperature without major problems and they are well-adapted to our climate. Home, roadside, pick-your-own, and large-scale commercial potential for southern peaches is excellent.

Soil and Site

Soil drainage, both surface and internal, is the single most important consideration when growing peaches

The Redskin peach, one of the finest peaches you can grow.

for home or commercial use. The ideal soil for peaches is 12 to 36 inches of sandy loam topsoil underlaid with a porous red clay subsoil. The topsoil and subsoil should have good internal drainage with no waterlogging problems. The clay subsoil should anchor the tree and serve as a major water reservoir. If yours is a silt or clay soil that has poor internal drainage (as determined by the drainage test on page 5), plant your peaches on raised beds to encourage maximum surface drainage.

The top of the raised bed, ridge, terrace, or berm should be at least 18 inches above the normal soil surface. If peaches are to be grown on lowlands, a drainage ditch will be needed in addition to raised beds. One should never underestimate the importance of internal soil drainage. It is the single most critical limiting factor in home peach production. Recent research by Dr. J. W. Worthington at Stephenville, Texas has shown the peach to be exceptionally drought hardy.

Location of a peach tree is important. On a cold, still March morning as the temperature drops to 29°F, a peach tree in full bloom located on a higher elevation with good air movement will suffer little or no freeze damage; but if it is in a low or flat area, the crop could be lost.

Good irrigation water should be available to deliver at least 15 gallons per tree per day when the tree is mature.

Varieties

There are over 500 peach varieties propagated, but very few are horticulturally important in the South. The following varieties have stood the test of time or have strong potential. They are listed in order of ripening, from the earliest ripening varieties to the latest. Numbers in parentheses indicate hours of chilling required.

EarliGrand: (250) clingstone variety; normally blooms in late January or February and produces moderately large, attractive fruit.

Springold: (700) clingstone variety; very heavy crops of very small red peaches.

Florida King: (400) clingstone; blooms early and ripens very early.

Bicentennial: (750) clingstone; heavy crops of medium-size fruit with adequate bacterial spot resistance.

Rio Grande: (450) semi-clingstone; regular crops of fair-quality fruit.

June Gold: (650) clingstone; heavy crops of high-quality peaches. Best variety for this chilling zone. Susceptible to bacterial spot.

Surecrop: (1,000) clingstone; heavy crops of large high-quality peaches.

Sam Houston: (500) large freestone; heavy crops of fair-to poor-quality fruit. Highly susceptible to bacterial spot.

Sentinel: (800) large semi-freestone; heavy crops of high-quality peaches. Tends to escape freezes and fruits when other varieties fail.

Red Haven: (950) large semi-freestone; regular crops of high-quality peaches.

Ranger: (900) medium-sized freestone; heavy crops of high-quality peaches.

Harvester: (750) large freestone; heavy crops of high-quality peaches. A firm, round variety adapted to mechanical harvesting.

La Faleciana: (600) freestone; produces heavy crops of high-quality, fuzzy peaches.

Redglobe: (850) large freestone; very high producer of high-quality peaches. Susceptible to bacterial spot.

Loring: (750) extremely large freestone; moderate crops due to frequent frost injury.

Milam: (700) medium-sized freestone; heavy crops of high-quality peaches.

Dixiland: (750) large freestone; heavy crops of excellent-quality peaches.

Redskin: (750) large freestone; heavy crops of excellent-quality peaches. Redskin is a seedling of Elberta and is one of the finest peaches grown in the South.

Frank: (750) medium-sized clingstone; used for "pickled" peaches.

White or *Honey* peaches have a 400-hour chilling requirement. The *Luttichau, Melba,* and *Pallas* varieties are commonly planted as dooryard trees.

Nectarines have not been grown with continual success in the South but new varieties such as *Sunrich, Sunreal,* and *Sungold* show promise for the future.

When selecting a variety for your area, it is important to understand that a chilling requirement is not a hard and fast rule; it is a general guide. A 600-hour variety can be grown in an 800- or 400-hour area; however, if the difference is greater than 200 hours, serious problems can result. If a variety does not receive sufficient chilling, it will not bloom and grow. If a variety's chilling requirement is satisfied by January 15, it can bloom before the last killing freeze. The best choice is a variety that requires slightly more chilling than your area normally receives. This would result in a slight delay in bloom which would hold the trees until the last freeze is past.

Spacing

A peach tree needs space to grow. You'll also need space to move your cultivator, harvesting wagon, and sprayer between the trees. Dry land orchards without irrigation are usually spaced 25 × 25 or 30 × 30. Many or-

Cut the nursery tree back to about 24 inches high.

Trim the roots to fit the hole.

Backfill the planting hole gently but firmly with soil . . .

. . . up to the point at which the base of the tree grew in the nursery row.

chards are now planted on 24-foot rows with 18 feet between trees. This spacing gives 100 trees per acre.

A 20 × 20-foot spacing is too close. But if your space is limited you can use drip irrigation and frequent pruning throughout the summer to keep the trees small. With this type of management, you can plant your peaches as close as 10 × 15 feet in a high-density spacing.

Training and Pruning

Peaches should be trained to form an open-center, vase-type tree. Cut the nursery tree back to 24 inches at planting. Wrap the lower 18 inches with a plastic tree guard or aluminum foil to prevent lateral shoots from

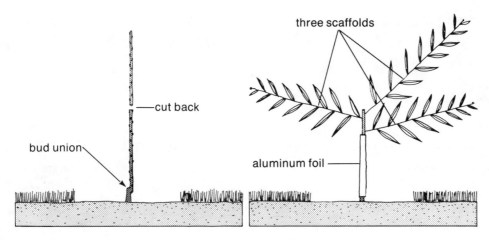

Figure 15. *Planting and summer training for first-year peaches and plums.*

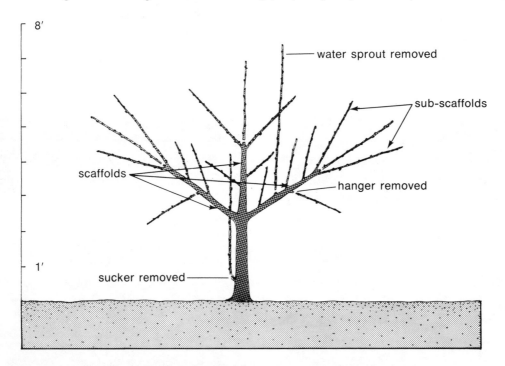

Figure 16. *Second or third dormant season peach pruning.*

developing and to protect the bark from contact with herbicides and rabbits. In June or July, leave five to six scaffold limbs near the top of the tree. This is the most important training in a tree's life. As the scaffolds develop 12 inches of growth, remove all but three shoots at or near the top (Figure 15). Cut the main scaffolds back to 32 inches the second year and select sub-scaffolds in June or July (Figure 16). Ideally, the tree will develop six sub-scaffolds; however, up to eight can be tolerated. Suckers and water sprouts should also be removed.

As the scaffolds and sub-scaffolds extend outward, they should form an angle of 45° or slightly lower. If the sub-scaffolds are more upright they will be far too vigor-

ous at the top of the tree. If they are lower, the limbs will hit the ground when they are loaded with ripe fruit. Small plastic newspaper bags filled with sand can be placed on fast-growing, upright sub-scaffolds to keep them at a 45° angle or slightly lower.

The mature peach tree should have approximately 40 percent of its wood cut out each year to stimulate new fruiting wood (Figure 17). All grey shoots that are two years of age or older must be removed, leaving only the fruitful 1-year-old red shoots on the six to eight sub-scaffold limbs. Hanger shoots which tend to grow down should be removed from the lower area of the tree; shoots which grow over 7 feet in the top of the tree

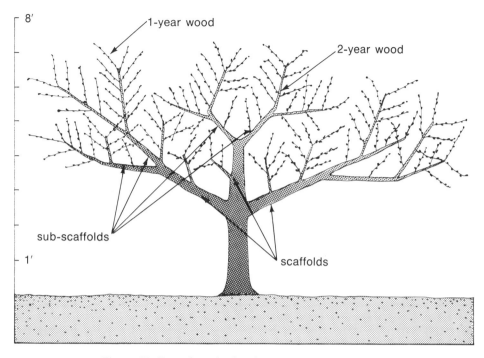

Figure 17. Properly trained and pruned mature peach tree.

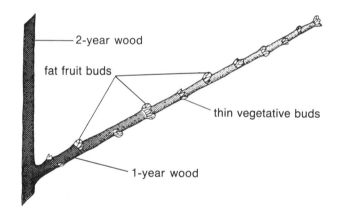

Figure 18. Dormant buds on one-year peach wood.

For good-sized, healthy peaches, thin out 70 percent of the fruit when they are less than 1/2" in diameter (or smaller on early-ripening varieties).

should be cut out. All shoots growing toward the center of the tree should be removed. Excessive growth in the fruiting zone should be thinned out to allow sunlight to enter. The center should be opened up every year. Again, suckers and water sprouts should be removed each year. The mature peach tree, when trained and pruned properly, should have six to eight sub-scaffold limbs with 50 red, 1-year fruiting shoots on each.

Fertilization

Peach trees must be pruned, irrigated, thinned, and cultivated for the best quality, yield, and optimum foli-

age. If you over-fertilize, too much growth can occur. Ideally, shoots on mature peach trees should grow 12 to 24 inches each year. This growth level will result in a heavy *fruit bud* set, Figure 18. If growth is greater than 24 inches, pruning will be difficult and too many *vegetative buds* will set. Limiting the fertilizer will control tree vigor and fruit bud set.

In February, apply 1 lb of 4-1-2 complete fertilizer to mature trees for each inch of trunk diameter at the

A properly thinned peach tree limb.

ground line. If good growth does not result, follow up with a 1-lb nitrogen fertilizer side dressing after harvest.

Young peach trees can be grown into large bearing trees in just 2 years with drip irrigation, summer pruning, herbicides, and frequent applications of small amounts of fertilizer.

Apply 1 lb of 4-1-2 complete fertilizer to 1-year-old trees in May. If the trees begin to grow rapidly, apply 1 lb of nitrogen fertilizer in June, July, and August. Repeat the same schedule the second year. Fertilize as mature trees the third year.

Thinning

Peaches tend to set an extremely large number of fruit each spring and unless you remove *70 percent* of it, you'll have difficulty obtaining good-sized fruit. Unthinned peach trees will not produce additional *pounds* of fruit, just more small fruit, and the general health of the tree will be endangered if the crop load is not reduced. Limb breakage, too, becomes a serious problem if unthinned.

Thin the small peaches to 6 inches apart when the fruit are less than ½ inch in diameter. Very early-ripening varieties such as EarliGrand, Springold, Surecrop, Bicentennial, and June Gold should be thinned during the bloom stage. Leave approximately 500–600 peaches on a mature healthy tree (see photo). A good rule of thumb: You cannot thin too much or too early.

Propagation

Nurserymen propagate peaches by two methods: June budding and dormant budding.

The rootstock seeds are stratified in moist refrigerated conditions from harvest to February. For June budding, a rootstock seedling is planted and grown from March until June, when the variety scion is T-budded onto the rootstock. The bark slips readily at this time. The graft will grow until first frost. These June-budded trees will grow from 16 to 48 inches in 4 months.

Dormant budded peach trees are not budded onto the rootstock until September. They are forced the following March and allowed to grow a full year in the nursery.

The nemaguard rootstock should always be used because peaches are subject to the rootstock nematode.

Peach Pests

Peaches are sprayed to prevent brown rot and control "cat-facing" insects such as stink bugs. Spray mature bearing trees with an insecticide and fungicide at least five times from March through July. A typical spray schedule includes sprays at pink bud, full bloom, petal fall, and shuck split followed by regular cover sprays 10 days apart. A dormant oil spray should be used to control scale insects. Your county Extension agent can supply you with a peach spray program specific to your area.

Brown rot is the major peach disease. It overwinters on the ground on old fungus-infested fruit called mummies. Spores move through the air to infest the blossoms first and fruit later. High humidity increases the problem greatly. It must be controlled by spraying to prevent development of the fungus on the blossoms and fruit. Starting early in the season is very important.

Brown rot on peaches. Prevent with early sprays of an approved fungicide.

Root knot nematodes can be a very serious problem on peaches. A nematode is a microscopic worm which feeds on young roots. Infected trees are weak, unproductive, and very sensitive to water stress. Fortunately, nemaguard rootstock can be used to prevent this nematode problem.

Stink bugs damage the peach when it is very small up until harvest. They suck juice from the fruit, and the wound leaves a very unattractive, pitted scar. Regular sprays with peach insecticides labeled for stink bugs will be needed for adequate control.

Peach tree borer was once a major limiting factor in peach culture. The insect infects the trunk, causing sap and jelly to ooze out at the ground line or just above it, finally killing the tree. A new chemical, chloropyrifos, gives excellent control if sprayed onto the infected trunks in mid-August. It must be sprayed when no fruit is on the tree or when all the fruit has been removed. Digging the borers out by hand or using moth ball crystals is no longer necessary for good control.

Bacterial stem canker is a very serious disease of southern peach trees. It is identified by jelly oozing out of buds, nodes, and wounds on young stems and limbs in August and September. The best control is optimum management and heavy fertilization. Do not prune from tree to tree without sterilizing the pruners with a 1:10 Clorox:water solution; otherwise you will spread the disease. Fall sprays of copper at leaf drop help reduce this problem.

Bacterial spot causes yellow leaves, premature defoliation, and deep spots on the fruit. It is best controlled with resistant varieties, optimum management, and a fall copper spray.

Scale insects will infect and kill the trees if they are not sprayed with dormant oil in February before bud break.

Harvesting

Peaches are non-climacteric fruit (if green, they won't ripen correctly after picking) and must be harvested when firm-ripe. If harvested before this stage, the fruit will not obtain its peak quality. Use three keys to determine if a peach is firm-ripe: it should have begun the final swell, yellow color will not have a green cast, and the fruit will "give" very slightly when squeezed in the palm of the hand. Harvest your trees at least four times to insure top size and quality.

Peaches will begin to bear a few fruit the second year. A healthy tree should produce a peck (¼ bushel) the third year and 1 bushel the fourth year. It should bear a full crop of 2 or more bushels the fifth year after planting.

PLUMS
Prunus salicina

Plums grow wild in 90 percent of the southern counties, and this is evidence of their adaptability. There are few tree crops which will require less attention and bear more fruit than the plum. While the peaches perform poorly as an individual tree, a single plum makes an excellent specimen tree.

Climate and Soil

The chilling requirements for plum varieties range from 400 to 1,000 hours. Plums can be grown in most southern soils: You can grow them in very sandy or clay soils where peaches won't survive. But for maximum growth and production, a sandy loam soil with good internal drainage is best.

Varieties that are primarily of Japanese origin are quite cold-sensitive and cannot be used in areas receiving temperatures below 0°F every year.

Varieties

Some plum varieties are widely adaptable and can be grown in essentially every area of the South. Others

The Morris plum, the best variety for the upper South. This high-chilling variety requires 800 hours of chilling.

have a rather high chilling requirement and are limited to areas receiving 800 or more hours below 45°F during the winter.

Morris is a large, beautiful, productive plum with bright blood-red skin and flesh. The plum is firm and high in sugar. The variety was selected as a Methley seedling by Dr. J. Benton Storey of Texas A&M University. The Morris plum has been described as the best plum variety for the upper South. It is a high-chilling variety and can be grown only in areas receiving 800 hours of chilling.

Methley is a medium-sized self-pollinating red plum with excellent quality and production. Methley is a relatively low-chilling variety and can be grown in areas receiving only 400 or 1,200 hours of chilling. Methley makes an excellent yard tree.

Ozark Premier is a very large, light-colored plum with excellent quality. It is a high-chilling variety, requiring 800 hours. The tree is moderately vigorous and quite productive.

Bruce is an extremely productive plum used primarily as a jelly variety since it has very little fresh fruit flavor. The Bruce is a self-sterile variety and must have the Methley as a pollinator. It has a low chilling requirement and can be grown in areas receiving only 400 to 1,000 hours of winter chilling.

Allred plum is a large tree that has red leaves, bark, and fruit. It makes a beautiful landscape specimen and also produces good fruit.

Producer and *Crimson* are new plum varieties from the Joe Norton breeding program at Auburn University. They are productive, delicious, and, more importantly, disease resistant.

Sweet Baby is a new plum developed in Georgia that looks outstanding at this time.

Robusto is a new, productive, high-quality plum that should be tested in the South.

Cherry Plums are excellent dwarf trees for areas receiving over 800 hours of chilling. The *Sapa* and *Hiawatha* varieties produce extremely high yields of small red plums on very small trees. The fruit should be thinned to prevent overcropping. Otherwise, the tree will fruit itself to death in only five or six years. These are excellent trees for those with limited space and cold winters.

Pollination

Most hybrids of the Japanese plum are self-sterile and require a pollinator. The Methley variety is an excellent

The Allred plum, an excellent specimen tree.

pollen-producing variety and should be in every planting. If you're only planting one plum tree, it should be a Methley.

Spacing

Commercial growers use 20 × 20 spacing, but high-density spacing at 12 × 24 or 10 × 15 feet will work well if you prune heavily each year.

Training and Pruning

Train plums to an open center, vase-type tree similar to the peach. The young tree should be trained during the first growing season by selecting three scaffold limbs 24 inches from the soil (Figure 15, page 28).

The plum will produce numerous long shoots from the scaffolds each year. Thin these out each winter, removing 20 to 30 percent of the shoots. Extremely vigorous, upright shoots should be headed back to no more than 8 feet from the soil (Figure 19). As the trees age, they will develop fruiting spurs that will bear the majority of the crop. Thin out the young, tall, vigorous shoots, and leave the shoots with fruiting spurs. If you don't prune plums each year, the tree will go into alternate bearing, producing an extremely heavy crop one year and no plums the next. The tree will then become exhausted and be very susceptible to bacterial stem canker.

Fruit Thinning

Plums respond beautifully to fruit thinning. Methley fruit can be increased from ½ inch in diameter to 1½

inches simply by thinning the crop load. The fruit should be thinned to 4 inches apart immediately after petal fall. Trees which go unthinned and overcrop will produce small-sized fruit and go into alternate bearing. Thinning is difficult but necessary.

Plum Pests

Plums don't require as rigid a spray program as peaches; however, you should spray to prevent brown rot and peach tree borer. You should also apply a dormant oil spray to control scale insects when they occur. Spray tree trunks and scaffold limbs with chloropyrifos in August to control the peach tree borer.

When purchasing new plum trees, be sure they're grafted on the nemaguard rootstock to protect the roots from root knot nematodes.

Bacterial stem canker is a serious problem in southern plums and peaches, and there is no good cure. Fortunately, some varieties appear to be less susceptible than others. The disease is first noticed as a jelly-like substance oozing out of the stems. In the dormant season, there will be lesions and scar tissue up and down the stem. These lesions should be thinned out before growth begins in March. Trees that are weak from overcropping, crowding, no thinning, and no pruning seem to become infested first. Therefore, it's important to purchase trees from a certified disease-free source and keep them healthy.

The Santa Rosa variety is highly susceptible to bacterial stem canker and is very short-lived in the warmer regions of the South; the disease is not as severe in the cooler areas of the South.

Plum varieties (from left to right): Methley, Allred, Bruce, Morris, and Ozark Premier.

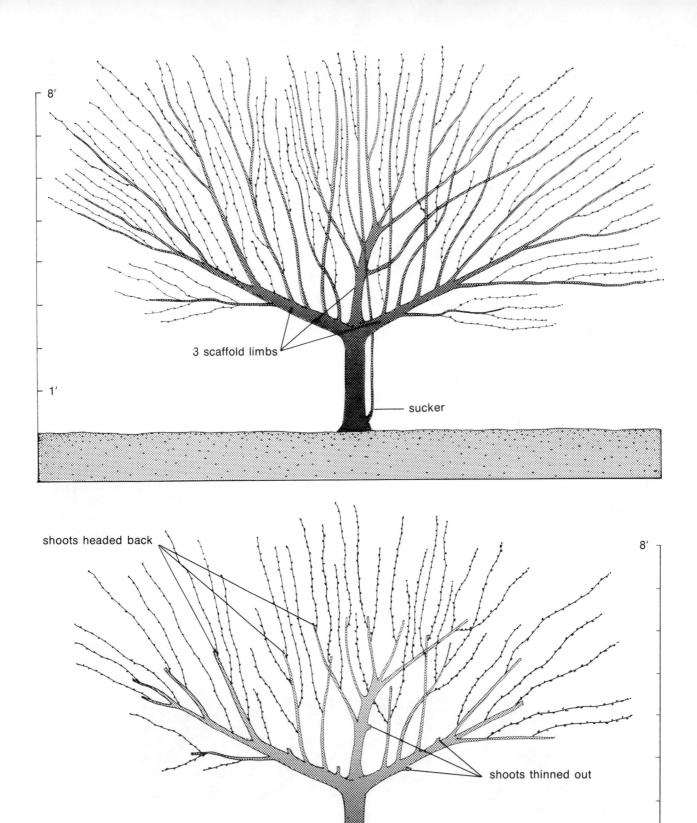

8'

1'

3 scaffold limbs

sucker

shoots headed back

8'

shoots thinned out

1'

sucker removed

Figure 19. A mature plum tree before and after dormant pruning.

Plum Curculio is a small weevil which affects plums, peaches, sour cherries, apricots, cherry plums, and nectarines. It attacks the fruit very soon after bloom by eating a hole and then laying its eggs inside the fruit. A worm hatches and feeds around the seed, producing a wormy fruit which usually ripens early and falls prematurely. It can be controlled with a regular spray program that includes an insecticide cleared for use.

Harvest

Plums are climacteric fruit and will ripen after they are picked. However, the plum must be showing some color before picking. The southern Bruce plum industry failed because the fruit was picked too early for proper ripening enroute to northern markets.

As with all fruits, *highest quality is obtained if the fruit ripens fully on the tree*. The taste of plums is excellent when eaten fully ripe straight from the tree. Southern plums also make delicious jelly.

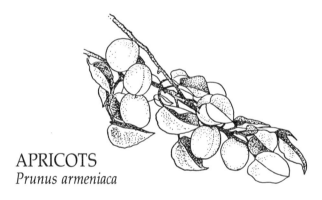

APRICOTS
Prunus armeniaca

It's extremely difficult to get apricots to fruit in the South because of their early blooming habit. A crop should be expected no more than every 3 or 4 years. So you should preserve a large percentage of the crop on the years the trees do produce.

Of the over 100 varieties evaluated in the South, the *Bryan, Hungarian, Bleinheim, Royal,* and *Moorpark* have performed best. Apricot trees are large, vigorous, attractive and require little pruning. Limbs should be occasionally thinned out to reduce the size of the tree. The fruit are highly susceptible to brown rot and will have to be protected with fungicide sprays.

SOUR CHERRIES
Prunus cerasus

The sour cherry is most certainly a cold climate fruit. We can only grow them in the northern limits of the South which receive over 1,200 hours of winter chilling. Winter sun scald injury is a common problem when they are grown here.

Montmorency is the primary variety propagated. The fruit is large, red, tart, white-fleshed, and is used almost exclusively in cherry pies. The tree is very large, productive, self-fruitful, and requires well-drained soil.

Young trees should be trained to a modified central leader like the apple or pear, and not like the peach. Once 4 to 6 scaffold limbs are selected on the central leader of the tree, only light pruning will be required.

Brown rot and leaf spot are common diseases of sour cherries and will need to be prevented with fungicide sprays.

Pome Fruits: Apples, Pears & Quince

APPLES
Malus pumila

Apples are a traditional northern crop, requiring 1,000 hours of winter chilling and cool fall nights for optimum fruit coloring and ripening. But they are grown, to a lesser degree, here in the South. North Carolina and Arkansas both have major commmercial apple industries.

Climate

Though you can grow apples in the South, remember that they are best suited for a colder climate. Trees live to be 50 years old and bear well in Oklahoma, Arkansas, Tennessee, North Carolina, and Arizona. When grown further South, they will ripen early and have poor red color. It is not uncommon to harvest apples in July in Georgia and Texas. The poor color is due to the absence of cool nights during the fall ripening period, and in the deep South it is possible to have no red color on some red varieties. In addition, warm southern temperatures can encourage severe fruit drop problems.

With this in mind, you'll be much happier with the performance of our southern apple varieties. An advantage for southern apples is that they ripen early and bring high prices. Though they have little red color, the taste is excellent. New varieties with strong red color are well-suited to the southern climate. Even more exciting, we now have some very low-chilling varieties that can be grown in areas receiving only 400 hours of chilling.

Soil

Apples can be grown in most soils, but they live and produce best in well-drained, fertile loam. Apples are extremely susceptible to root rot and should not be grown in soils with a history of any root rot problems.

Varieties

Apple varieties are of two types, standard and spur. Standard-type trees are large and require longer to come into production. Spur-type trees are smaller and bear at an earlier age. Apple rootstocks, more than in other fruits, will influence variety performance.

Starkrimson Delicious is a high-yielding red spur-type with a typical delicious shape. It has proven to be one of the better red spur apples for the South. It will develop good red color in the cooler areas and can be grown in areas receiving 700 hours of chilling.

Golden Delicious is a high-producing yellow standard-type apple with good quality. The skin is slightly russetted. Golden Delicious is a high-chilling variety and cannot be grown satisfactorily in areas receiving less than 900 hours of chilling.

Mollies Delicious is a standard-type red delicious apple which produces well in the South. It does not develop good red color in the Gulf Coast area, but can be grown in areas receiving only 600 hours of chilling. It has a very good taste.

Jersey Mac is a red standard-type variety. It is a high-chilling variety and should be grown in the 1,000-hour areas. Jersey Mac is similar to the McIntosh variety. It is very susceptible to fire blight. *Jersey Mac* has excellent

The Starkrimson Delicious apple, a high-yielding variety with good red color in cooler areas of the South.

red color while other varieties have serious red color problems.

Granny Smith is a green apple when ripe, and it is a good one for homeowners. It is reported to have a wide chilling range, and it stores well on the tree. Since it is relatively new to the South, its real potential is unproven.

Gala is a new yellow variety which can have a red blush. Its taste is outstanding. It is a heavy-bearing standard-type that requires 700 hours of chilling.

Anna is a low-chilling red standard-type variety, which seldom develops red color. It produces large-sized fruit, which ripen very early and have a very limited shelf life.

Fugi is a new variety being tested across the South. It is very sweet and delicious but rather unattractive and late.

Dorsett Golden is a low-chilling yellow standard-type variety, which produces well in combination with Anna. It can have a pink blush and good taste. The fruit is firm and sweet.

Aldina is a new low-chilling red standard-type variety for the 400- to 600-hour zone. It has large fruit which ripen early. It is a good producer and has a unique cinnamon flavor.

Pollinators

Golden Delicious is a good pollinator, as are King David, Winter Banana, and Dorsett Golden.

Cross-pollination is very important in apples. If not essential for some varieties, it will improve fruit set in *all* varieties. Different varieties should be planted, with at least one tree out of ten as a pollinator. Bees are very important at full bloom, so don't use insecticides during this period. Cultivated bee hives will definitely improve fruit set if placed in the orchard during full bloom. One or two hives should be used to each acre of apples.

Rootstocks

Apple rootstocks can influence the size of the tree, the year it comes into production, the strength of the tree, and its susceptibility to root problems. There are two types of rootstocks: seedlings and clonal. The seedlings are grown from seed and make a large tree which is considered standard. Seedling rootstocks are not recommended in the South because of their susceptibility to the woolly apple aphid. This insect feeds on the roots of apple trees and eventually kills them.

Clonal rootstocks tend to dwarf the trees and bring them into production earlier. Clonal rootstock trees are grown from stem cuttings rather than seeds; thus each individual clonal rootstock is genetically identical. Most of the clonal rootstocks were collected and evaluated at the East Malling Research Station in Kent, England and carry the "M" indicator and a number for identification.

M-111 is the clonal rootstock used for spur-type varieties. It will produce a semi-dwarf tree which is approximately 85 percent the size of a seedling tree. M-111 has demonstrated good root rot tolerance in the South and is relatively long-lived. It is resistant to the woolly apple aphid. The trees on M-111 will come into production earlier than trees on a seedling rootstock.

Using wooden spacers to spread scaffold limbs on immature apple trees.

M-106 is the clonal rootstock to be used for standard-type varieties, which normally make larger trees. With M-106 as a rootstock, these varieties will be semi-dwarf and produce a tree approximately 70 percent the size of a seedling tree. M-106 will bring the tree into production at an earlier age than M-111 or seedling rootstocks. It is well adapted to sandy soils, drought-tolerant, anchors the tree well, and does not produce excessive suckers. M-106 appears to have less root rot resistance than M-111. It is resistant to the woolly apple aphid.

M-26 and *M-9* are fully dwarfing clonal rootstocks which can be used to produce very compact trees. The soil should be pretreated with a chemical sterilant because they do not have good root rot resistance. They will bear apples even when grown in containers. They are very susceptible to fire blight; consequently, suckers should be removed as they appear. Trees grafted onto M-26 and M-9 will begin bearing at a very early age.

Spacing

The new clonal rootstocks and spur-type varieties allow closer spacing than the traditional 30 × 30 apple spacing. Depending on variety type and rootstock, apples can be spaced 10 × 20, 15 × 20, 12 1/2 × 25, or 20 × 25. The closer spacings are used when the spur-type varieties are grafted onto M-106. Homeowners can experiment with M-26 and M-9 rootstock as close as 5 feet apart.

Training and Pruning

Train your apples to a modified central leader by careful training at the end of the second, third, and fourth growing seasons. Cut the tree back one-half at planting and allow it to grow freely the first year.

At the end of the first growing season, select a central leader and cut it back. Remove other vigorous shoots in the center of the tree so that only one central leader remains. Two or three scaffolds can be selected; all other shoots should be thinned out (Figure 20). Clothespins can be used as spacers while the scaffolds are small.

At the end of the second, third, and fourth growing seasons the central leader should again be selected and cut back. Each year, select scaffold limbs that are not directly above another scaffold limb. All other lateral shoots should be thinned out. Spread the scaffold with plastic bags filled with sand or a spacer (Figure 21). The scaffold limbs should be weighted or spaced so that they are slightly lower than a 45° angle. If the limb moves above the 45° line, use a larger spacer or move the sand bag further out on the limb. There are several types of spacers you can use (Figure 22). Several lengths, from 6 to 24 inches, should be made or purchased. As the limbs increase in size and begin to bear, move the spacers to newer limbs up the tree. Water sprouts will develop on the spread scaffold limbs. They will need to be pinch-pruned two to three times during the growing season to retard their growth (Figure 23). If they grow over 12 inches, remove them.

Mature apples will require little pruning except occasional thinning out or to remove fire-blight-infested shoots.

Fertilizer

Several 1/2-lb applications of 3-1-2 or 4-1-2 complete fertilizer should be made each year for 3 years after

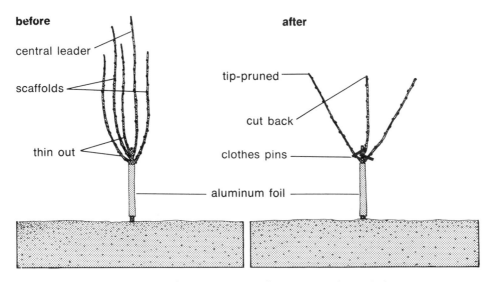

Figure 20. First dormant season apple pruning (before and after).

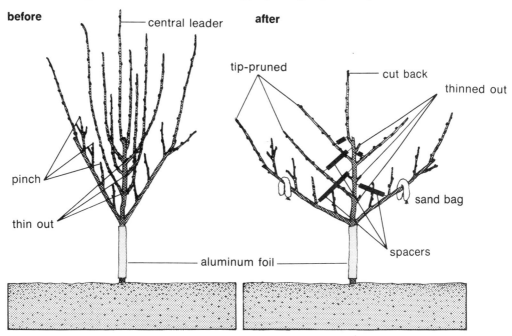

Figure 21. Second dormant season apple pruning (before and after).

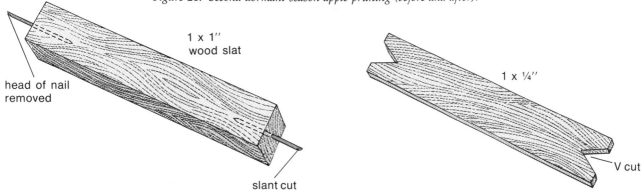

Figure 22. Wooden spacers for spreading young scaffold limbs on apple trees.

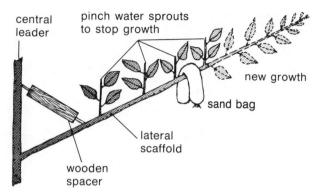

Figure 23. Pinch water sprouts on young apple scaffold limbs to keep them less than 12" long.

planting. Avoid making one heavy application. Spread the fertilizer out to prevent dehydration burn.

Mature apple trees should not be forced into excessively vigorous growth with nitrogen fertilizer. If the trees are making less than 6 inches of terminal growth annually, apply 1 lb of a complete fertilizer for each inch of trunk diameter before growth begins in the spring.

Apple Pests

Southern apples are attacked by quite an array of pests. A regular spray program is necessary for the production of clean fruit. Contact your county Extension agent for an apple spray schedule for your area.

Mites are a major problem. Mite damage can be recognized by a rusty or dusty appearance on the leaf surface. The mite is very small but can be seen on the underside of the leaf with a 10X hand lens. Mite populations increase rapidly and seem to invade the apple trees overnight. Control with miticide sprays cleared for use on apples.

Apple scab is a problem in high rainfall areas. It appears on the leaves as small spots which result in premature leaf drop and reduce tree vigor. The fruit are affected by dark brown, corky spots. The disease is a fungus and is controlled with fungicide sprays.

Bitter rot is a common apple problem in the South. It appears first as a small black spot which continues to increase in size. Eventually the entire fruit can become consumed and will shrivel to a small, black, wrinkled, mummified fruit. All diseased or mummified fruit should be collected and burned in the dormant season. A fungicide spray cleared for bitter rot control on apples should be satisfactory.

Fire blight as described for pears (page 43) is also a major problem in apples, though usually not as severe.

Harvest

Apples are climacteric fruit and can be harvested before they are tree ripe. Once the desirable size is reached the fruit can be picked. Because of our warm southern climate, the color will not be outstanding. But you can spray new ripening chemical, Ethrel, onto the fruit a week before normal harvest to help improve fruit color. A stop drop chemical called Fruitone T must be included with the Ethrel spray to prevent premature fruit drop.

SOUTHERN PEARS
Pyrus pyrifolia, hybrids

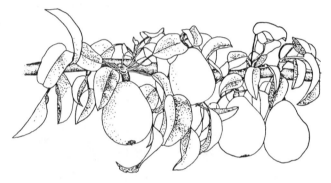

The southern pear is an excellent home fruit tree for the South. It is extremely well adapted and some varieties are very resistant to fire blight. Very old oriental pear trees can be found throughout the South as landmarks of old homesites. The tree fits well in the landscape and produces a beautiful show of white blossoms in the spring. The European pear, *Pyrus communis*, which is common in the North and a commercial variety on the West Coast, cannot be grown in the South because of its susceptibility to fire blight. The Southern pears are different from the newly introduced Asian pears (page 44).

Climate and Soil

Pears can be grown in a wide variety of soils. They can grow in heavy clays or light sands. Extremely fertile soils can stimulate too much new growth, which is a problem.

High humidity and rainfall encourage and spread fire blight. Several oriental varieties have a relatively low chilling requirement, some as low as 400 hours. Others will grow well in areas receiving only 200 frost-free days. Few crops are better adapted to the southern climate than the oriental pear hybrids.

Orient pear, more resistant to fire blight than any other southern pear. The tree bears fruit after 4 years.

Varieties

Fire blight resistance or tolerance is the major criterion when selecting pear varieties for the South. Factors such as fruit quality, production, and tree strength must be considered.

Orient is a large, early, russetted pear with excellent flesh quality. The tree has the highest fire blight resistance of any southern pear. It is strong and very easy to train. Orient holds its leaves well in the fall and is a regular producer. The chilling requirement is relatively low; it can be grown in areas receiving 400 to 1,200 hours of chilling. Orient is reported to be self-sterile, but seems to have no problem fruiting when a pollinator is not near. The tree will bear fruit in approximately 4 years.

Ayers is a small early pear with excellent flesh and canning quality. It's the best fresh fruit pear southerners can grow. The fruit is yellow with an attractive red blush and has few grit cells. It is reported to be self-sterile. Ayers has relatively good fire blight resistance. Its chilling requirement is relatively high and it should not be grown in areas receiving less than 700 hours of chilling. It requires at least seven years to bear fruit.

Kieffer is a fair-quality hard pear which is highly tolerant of fire blight. The disease does not spread rapidly throughout the tree as in other varieties. The fruit has a moderate amount of grit cells and lacks the typical pear flavor. Kieffer trees are vigorous, very productive, heat-tolerant, and bear in approximately 5 years. This variety has a relatively low chilling requirement and can be grown in areas receiving 400 to 1,200 hours of chilling. Kieffer is reported to be self-sterile.

Leconte is the highest-quality old oriental hybrid pear grown in the South. It doesn't have excellent fire blight resistance and a tree can be killed in one season. The fruit has a typical pear shape and can be eaten fresh. It makes excellent preserves and has a relatively low grit cell content. The tree has a low chilling requirement but can be grown equally well in the 1,200 hour chilling zone. LeConte is reported to be self-sterile.

Moonglow is a quality oriental pear hybrid with good fire blight resistance. The tree is vigorous, upright, and bears in approximately 6 years. The fruit has excellent flesh and good canning and preserves quality. Moonglow has viable pollen and is an excellent pollinator. It has a relatively high chilling requirement and should not be grown in areas receiving less than 700 hours of chilling.

Garber is a fair-quality early pear shaped like an apple. It has good fire blight resistance and a relatively low chilling requirement. Garber has a moderate amount of grit cells and is an excellent canning or preserves pear. The tree is vigorous but has a tendency to lose its leaves in the fall.

Maxine or *Starkings Delicious* is a good eating pear. It is a mid-season variety and is low in grit cells. The tree is vigorous, productive, and has a relatively high chilling requirement.

Baldwin is a good variety for the Gulf Coast, where fire blight resistance is extremely important. The fruit is average-sized and of fair quality.

Rootstocks

Southern pears should be grafted onto *Pyrus calleryana* rootstock. This rootstock is adapted to a wide range of conditions and has a high degree of resistance to fire blight. Many homeowners purchase a pear tree as a grafted variety and later learn that they have the *Pyrus calleryana* rootstock. The fruit is very small with 4 to 10 fruit in a cluster. The tree makes a beautiful yard tree and produces very attractive blossoms in the spring. Several outstanding clones, such as *Bradford* and *Aristocrat*, have been propagated exclusively for ornamental use.

Pyrus calleryana, the rootstock on which all southern pears should be grafted.

Spacing

Pears become large trees and should be spaced at least 25 feet apart. They are not adapted to high-density spacing because they cannot be pruned severely. Pears grow best under mowed sod, without cultivation or mulching.

Training

Young pear trees should be trained to a modified central leader system similar to the apple and pecan, with 4 or 5 scaffold limbs at a 45° angle or lower. The tree should be cut back one-half at planting, Figure 24. At the end of the first growing season select a central leader and again cut it back one-half. Thin out the other very vigorous shoots that are competing with or are equal to the size of the central leader. Pinch-prune laterals on the lower trunk to inhibit their growth. These short lateral shoots are for trunk productions and food manufacturing.

At the end of the second growing season the first scaffold limbs can be selected 36 to 48 inches above the soil. Choose three or four limbs that aren't directly above another (Figure 24). A symmetrical spiral staircase pattern is ideal but not always possible. Remove the short lateral shoots on the lower trunk and select a vigorous upright shoot as the central leader and again cut it back one-half. Other vigorous shoots at the previous year's cutback point should be thinned-out. Head back the selected scaffolds one-quarter.

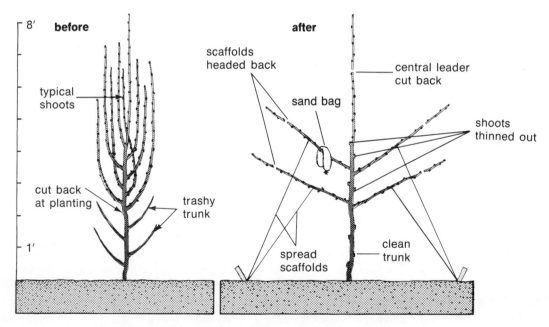

Figure 24. A two-year-old pear tree before and after pruning.

The scaffold limbs will grow straight up if you don't spread them at this point. This can be done with twine or cord pulling the scaffold limbs down (Figure 24). Water hose pieces, leather, cloth, or nylon stockings can be used to protect the scaffold limb where the string is attached. The latest method of spreading scaffolds is plastic bags filled with sand placed over the scaffold to weight it down to less than a 45° angle. Scaffolds should be selected, spread, and headed back for the final time after the third year's growth.

Summer Pruning. Vigorous vertical sub-scaffold shoots will begin to develop along the spread scaffold limbs. These should be thinned out to only two or three per scaffold. These vertical sub-scaffold shoots will develop fruit-bearing spurs the next season. Should a vertical sub-scaffold become severely infected with fire blight, remove the entire sub-scaffold. Consequently, your tree should have eight to twelve sub-scaffolds should a major fire blight infection occur.

Always thin out fire-blight-infested shoots as soon as they develop (photo).

Winter Pruning. Once the scaffold and sub-scaffold limbs are selected, very little winter pruning will be needed. You'll only have to thin out fire-blight-infested shoots and water sprouts. When the tree is mature and its growth rate has reduced to less than 6 inches a year, alternate bearing (fruit every other year) can occur and winter pruning can be used to stimulate new growth. Don't thin out more than 20 percent of the bearing wood; heavy pruning stimulates excessive growth and the tree will become susceptible to fire blight. Even more important, mature pears should never be headed back severely or dehorned, for this too stimulates excessive growth.

Fertilizer

Pear trees should be maintained in a low vigor state to reduce the possibility of fire blight. One lb of 3-1-2 complete fertilizer can be applied to young trees in February or March. However, if the vegetative shoots grow more than 24 inches in 1 year, do not fertilize.

Mature bearing trees should make less than 6 inches of terminal growth annually. If they don't, apply 1 lb of 3-1-4 complete fertilizer for each inch of trunk diameter. Don't use manure or other organic slow-release fertilizers; they can stimulate succulent growth susceptible to fire blight.

Pear Pests

Fire blight is by far the most serious pest with southern pears. Proper variety selection and cultural practices are essential in helping prevent fire blight. This bacterial disease affects the blossoms, and later the stems. Bees vector the bacteria to the blossoms and rain washes the bacteria throughout the tree. Occasionally, entire blossom crops are lost to the disease. Fire blight is easily recognized by black leaves (photo).

A pear tree with a fire blight-infested limb (lower right).

Shoots with black leaves should be thinned out as soon as they appear. If they aren't, the bacteria will be washed onto larger stems and limbs, where new infection will occur. New fire blight lesions must be thinned out in winter pruning or the tree will die.

Actinomicin and copper sprays are labeled for fire blight control, but they don't do a thorough job unless they are applied very frequently and over a long period of time. These sprays should only be used for preventing or controlling fire blight in the bloom stage.

Leaf spots due to fungi in the spring and late summer can cause premature defoliation of pear trees, and defoliation can cause severe tree stress if the tree is heavily loaded with fruit or if drought follows the infestation. After defoliation, late summer heat can satisfy the tree's chilling requirement. This will result in a full bloom in September or October, which is a very undesirable situation, since the fall bloom will stress the tree's food supply. Most fungicides cleared for pears will prevent these leaf spots if used when the spots first appear. The diseased leaves should be plowed under or swept and burned in the winter months. Good management also reduces the problem.

Harvesting and Ripening

Southern pears are climacteric and may be harvested hard and ripened under special conditions. Ayers, Moonglow, LeConte, and Maxine will ripen on the shelf several days after picking. Orient and Garber will require about a week. Kieffer should be individually wrapped in paper and held at room temperature for approximately 14 to 30 days for best flavor.

ASIAN PEARS
Pyrus pyrifolia

Asian Pears have been propagated and grown commercially in China, Japan, and Korea for hundreds of years, but have only recently become a popular crop in California. As Asian pears reached our local produce departments, southerners became interested in trying this new fruit.

The first commercial orchards of the South are just now being planted, and it is too early to determine which varieties will perform best in our hot, humid climate. Fire blight is a serious problem at bloom time in California and will probably be more serious in the South. Trees will need to be trained to numerous vertical scaffold limbs, so that, should one scaffold limb become infested with fire blight, it could be removed. Antibiotic sprays could also be very helpful during full bloom.

The culture of Asian pears will be basically the same as Southern pears. The most popular varieties today are *20th Century* and *New Century*; however, numerous other varieties such as *Shinko, Seuri,* and *Hosui* are being tested because of their resistance to fire blight. The fruit are uniquely different from Southern pears in that the flesh is crisp to crunchy and highly aromatic. The fruit ripens in late June and July and is extremely delicate, requiring very gentle handling. At the market they are individually wrapped in foam diapers to prevent bruising the flesh.

The shapes and skin color of Asian pears vary greatly. The skin color can range from the very smooth, brilliant yellow of *20th Century* to the orange russet of *Seuri*. *Pyrus calleryana* is the best rootstock for the South. A low-chilling line of Asian pears is recommended south of the 600-hour chilling zone. They include the *Ya Li* and *Tsu Li*. Cross-pollination is beneficial and more than one variety is recommended. Only time and experience will determine the future for Asian pears in the South.

QUINCE
Cydonia oblonga

The quince is a small tree which produces a rough pear-like fruit on short spurs or the end of shoots. It can be grown in all areas of the South. The fruit is not eaten fresh, but it can be used for preserves or jelly. The quince is commonly used as a dwarfing rootstock for pears.

Berries: Blackberries, Blueberries & Strawberries

BLACKBERRIES
Rubus sp.

Blackberries are a truly southern crop. Wild dewberries grow along every railroad track and fence row in the South, testifying to their adaptability. And for cultivation there are new varieties of very productive, high-quality blackberries.

Climate and Soil

Blackberries can be grown in the 1,200- or 100-hour chilling zones. They can withstand extreme heat and summer drought, both of which are common in the South. Few crops are better adapted to our southern climate than blackberries.

Blackberries will grow in almost every southern soil as long as it drains well. They'll grow in soils ranging in pH from 5.5 to 7.5; chelated iron will have to be added to highly alkaline soil to prevent iron chlorosis. The South's commercial blackberry industries are located in areas of deep, sandy, moderately acidic soil.

Varieties

Two basic types of brambles grow in the South: the erect blackberry and the trailing southern dewberry. The erect blackberry is very productive and does not require trellising; dewberries abound in the wild and are seldom planted. However, should you grow dewberries, they will require a trellis. (See page 84 for information on wild blackberries and dewberries.)

Brazos has been the South's leading erect blackberry for over 30 years. It has a high yield, large fruit, insect and disease resistance, and drought tolerance. The fruit is very attractive but has a very acidic taste. Brazos berries go beautifully with cream and sugar for dessert and make excellent pies, cobblers, and jelly.

Cheyenne and *Choctaw* are very erect blackberries developed by the University of Arkansas for mechanical harvesting. They are also outstanding as home varieties, with high yields and firm, attractive fruit.

Humble is an old erect blackberry variety which produces moderate crops of small, sweet berries. It is very important because it is the only major variety resistant to the serious double blossom disease. Humble is becoming difficult to purchase, which is unfortunate be-

The erect blackberry, an easy crop to grow.

cause it is the very best blackberry along the Gulf Coast where double blossom is severe.

Brison, Womack, and *Rosborough* are new erect blackberry varieties released by Texas A&M University. All have higher yields and higher quality berries than the Brazos. Brison has produced well in clay soil along the Gulf Coast. Womack has produced well in the 1,000-hour chilling zone in sandy soil. Rosborough has produced well in clay and sand.

Navahoe is a new thornless blackberry for the South, developed by Jim Moore at the University of Arkansas. It offers a great deal of promise, and appears to be superior to all of the previously tested thornless blackberries.

Hass is a new USDA thornless blackberry, which has excellent quality but low yields. It will perform best if tied to a trellis. It performs best in the area above the 600-hour chilling line.

Austin Dewberry is a typical-flavored dewberry produced on trailing canes. Fruit size and yield are much less than those of the recommended erect blackberries.

Varieties to Avoid. The following blackberries are *not* recommended because of their low yields, cultural difficulty, disease susceptibility, or weak canes: Lawton, Young, Barpen, Fluit, Gem, and old thornless varieties.

Propagation

Blackberries are one of the easiest fruits to propagate. Simply dig root cuttings the size of a ball-point pen and transplant to a new location (Figure 25). The roots should be dug in late winter before growth begins and stored in cool, moist sawdust to prevent drying out. Cuttings are usually bundled in packs of 10 to 50 roots. Don't forget to label each variety. They can also be stored moist in polyethylene bags and refrigerated at approximately 45°F.

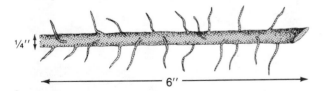

Figure 25. A blackberry root cutting.

Planting and Spacing

Plant blackberry root cuttings horizontally about 2 inches below the surface in heavy clay soil and 4 inches in lighter sandy soil. The planting bed should be well

Flowering Brazos blackberries.

cultivated several times in the winter to kill all permanent weeds.

Blackberries are easy to over-plant. They are very productive and are quite difficult to prune so don't plant more than you can handle. Plant 3 feet apart in rows or set in hills 10 to 20 feet apart (Figure 26). If you use the hill system, plant at least three root cuttings per hill. If you use the row system, plant one root cutting every 3 feet.

Hand pruning is easier with the hill system, as is weed control through cross cultivation. This system is also especially adapted to hand harvesting in a pick-your-own operation. The only major disadvantage is that the hill system requires considerable space. The row system will produce much higher yields in less space.

Pruning

Erect blackberries must be pruned annually for healthy plants and regular production. If they aren't, disease and insect problems will soon kill the canes. Erect blackberries are biennial plants, producing non-fruiting prima canes the first year and fruiting canes the following year.

The canes will die after fruiting and should be removed (Figure 27). A blackberry hill will always have 1-year-old non-fruiting prima canes and 2-year-old fruiting canes (Figure 27). Cut back the prima canes when they reach 36 to 48 inches to encourage lateral shoot formation. Remove fruiting canes immediately after harvest. You'll need gloves and lopers to protect your hands from thorns.

In the more southern areas that receive 300 or more frost-free days, the entire row or hill can be mowed

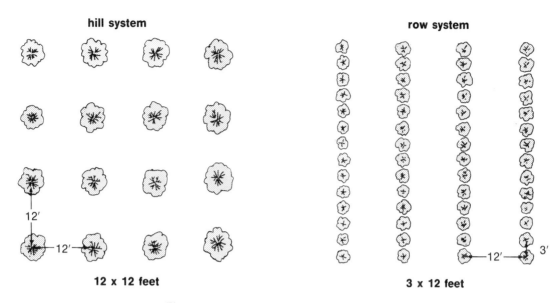

Figure 26. Erect blackberry planting systems.

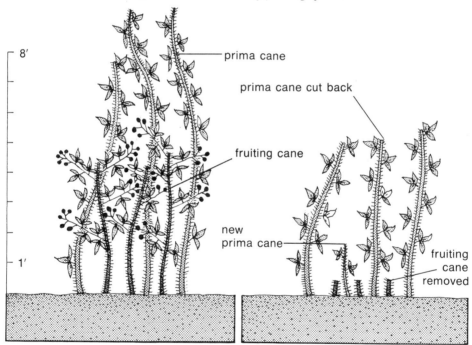

Figure 27. Blackberry pruning immediately after harvest.

down immediately after harvest with a heavy-duty lawn mower. Sufficient prima canes can be grown before frost for next year's production. Drip irrigation and fertilizer will have to be applied to obtain the amount of growth required after mowing.

weeds such as Bermuda grass and Johnson grass with contact herbicides. Pre-emergence herbicides can be used to prevent annual weed seeds from germinating. The new grass-selective contact herbicides can be used effectively on blackberries.

Weed Control

Blackberries are very shallow-rooted and should not receive deep cultivation. Before planting, kill perennial

Fertilizer

Apply 1/4 lb of 4-1-2 complete fertilizer per plant in February or March before growth begins. Apply the same rate of a nitrogen fertilizer immediately after fruiting and pruning.

Blackberry Pests

Insect and disease pests can be reduced significantly if you follow a good pruning and weed-control program. If you let dead or diseased canes accumulate and form a tangled mass, problems are sure to arise. This is the main reason you should keep your planting small.

Double blossom or *rosette* is a serious disease of all varieties of southern erect blackberries except Humble. Canes with the disease will produce masses of leaves in a rosette or witch's broom-type cluster. The blossoms will grow crazily, sometimes having leaves inside the flower. The disease is caused by a fungus, which enters the young prima canes from spores produced in the flowers. It can be inhibited or prevented if you prune out all new prima canes immediately after bloom. This is not a major problem on small plantings. Benomyl fungicides can also be sprayed onto the plants and blossoms at full bloom to help reduce double blossom.

Inspect fruiting canes in February; if rosetted, you'll need to thin them out before growth begins in the spring. In severe cases, mow down all the canes in May to prevent infection next year.

Anthracnose is another common disease of blackberries. The fungus can be identified on the canes as purplish spots with gray centers. The spots are usually sunken slightly. Anthracnose can be controlled with fungicide sprays from late winter continued to 40 days before harvest or early April. Liquid lime sulfur sprayed at or just before bud-break can also reduce the problem.

Red neck cane borer is an insect that feeds on the canes near the ground line. The cane will swell about 50 percent 2 to 4 inches above the ground. The best control is good sanitation and removal of infested canes as they appear.

Harvest

Blackberries are one of the earliest fruit crops in the South, ripening in late May and early June. The newer varieties will produce at least 1/2 gallon per foot of row or 2 to 3 tons per acre. Blackberries are not climacteric and should be harvested fully ripe. Under normal conditions a variety will bear fruit for approximately 2 weeks. Don't let the ripe fruit sit in the sun; place it in a refrigerator as soon as possible.

RABBITEYE BLUEBERRIES
Vaccinium ashei

Rabbiteye blueberries are growing in popularity all across the South. They grow wild along streams in southern Alabama, southern Georgia, and northern Florida. The first cultured plantings were in northern Florida at the turn of the century; extensive development has only come recently.

Part of the reason for the popularity of rabbiteye blueberries may be that no major pests have been identified on them. As increasing numbers of northerners move south, the demand for blueberries will increase. Southerners have not yet discovered the outstanding quality of rabbiteye blueberries. They associate them with wild huckleberries and sparkleberries, which are in the *Vaccinium* family but are characteristically small and tasteless.

Soil and Climate

Rabbiteye blueberries are one of the few crops which will require very special soil: a pH of 4.0 to 5.0 is re-

The Tifblue rabbiteye blueberry is more cold hardy than most other southern varieties. Its delicious fruit is large and light blue in color.

quired for good plant growth; the plants will not live in soils with a pH above 5.5.

The plants' feeder roots are very close to the surface and do not have root hairs; therefore, good soil moisture management and heavy mulches will be needed. Deep sandy soils, which tend to dry out, cannot be used for rabbiteye blueberries unless they are drip irrigated. Unmodified heavy clay soils with poor aeration and little internal drainage will not do.

A native southern plant, the rabbiteye blueberry can be grown in all areas south of the 1,000-hour chilling zone. They cannot be grown in areas that frequently receive temperatures below 0°F because winter damage will occur. Folks in colder areas should grow the northern high-bush blueberry; the rabbiteye blueberry is much more heat-tolerant than the northern high-bush blueberry.

Varieties

In the wild the rabbiteye blueberry will vary considerably in bush size, number of suckers, and fruit quality. Most cultivated varieties are mature in 10 years and will be 15 feet tall and 10 feet wide. The bush will consist of numerous suckers, which develop from the crown area. The fruit are borne in the top of the bush on shoots which grew the year before. The berries have a typical blueberry shape, a delicious taste, and a good sugar content.

Woodard is a very large, light blue rabbiteye blueberry that ripens early. The bush is moderately vigorous and the fruit are very high-quality. Woodard is a must variety in any planting.

Tifblue is the most universally outstanding rabbiteye blueberry grown to date. The fruit are large, light blue, and ripen late in the season. The bush is vigorous and very productive. Tifblue is more cold hardy than most rabbiteye varieties. It should be the predominant variety in any planting.

Garden Blue produces a very small, light blue midseason fruit. The bush is moderately large.

Delite is a showy variety that is good as an ornamental. The fruit are small and light blue (red and pink when immature).

Hombell produces a large fruit that ripens in midseason. It produces moderate crops on vigorous plants.

Menditoo is a large fruit that ripens late in the season. The bush is moderately productive and vigorous.

Briteblue is a new, moderately vigorous plant, which produces firm, large, light blue berries in moderate to heavy crops. The berries have a waxy bloom and should not be harvested before they are fully ripe. It normally ripens in early to midseason.

Climax is a new rabbiteye blueberry, which is early in ripening. Most of the fruit ripens in a short period of time. The crop load is moderate to high.

Brightwell is a vigorous, upright plant, which produces outstanding yields of medium-sized fruit. It ripens early to midseason.

Sharpblue is a new variety from the Wayne Sherman program at the University of Florida. It is a low-chiller for areas receiving 600 hours or less of temperatures below 45°.

Pollination

Rabbiteye blueberries benefit from cross-pollination. At least three varieties should be in every planting. All varieties seem to pollinate each other. Bees are extremely important in carrying the pollen from flower to flower. Small commercial growers should have beekeepers place hives in their planting during full bloom to insure good cross-pollination.

Spacing and Planting

Plots for rabbiteye blueberry plantings should be treated with glyphosate herbicide and well-tilled 3 months before planting to kill all weeds. In low, flat areas the beds should be raised to aid surface drainage.

Work organic matter as thoroughly and as deeply as possible into the planting spot prior to planting. It's difficult to add too much organic matter to the planting spot. (Peat moss is the best source of organic matter.) Ideally, till 1/4 to 1/2 bushel of organic matter into the soil for each plant prior to planting.

Two-year-old transplants give the best growth. Nurseries propagate them from small stem cuttings. Purchase bare-root or container plants from a well-known nurseryman or order directly from an exclusive rabbiteye blueberry nursery. Make certain the bare roots do not dry out, and, where container plants are used, separate the roots from the container ball when planting

Plant your rabbiteye blueberries 1 inch deeper than they grew in the nursery row. Also, cut the tops back one-half at planting to balance the tops with the roots. Set the plants on rows 10 to 12 feet apart; space plants no closer than 6 feet apart unless you want a hedge for limited space. The plants will crowd when mature if spaced closer than 6 feet.

Fertilizer and Mulch

Rabbiteye blueberries are very sensitive to commercial fertilizer. Use only ammonium sulfate or special azalea

A thick layer of mulch to acidify and cool the soil, and to hold moisture in, is essential for healthy rabbiteye blueberries. Mulch is especially important for young plants.

or camellia fertilizers. These should be used in frequent, very small applications rather than one heavy application. Don't use the nitrate-type fertilizers—they can kill the plant—and don't apply any commercial fertilizer the year the plants are set. Apply 1 oz of ammonium sulfate the second year. The rate can then be increased 1 oz per year but shouldn't exceed ½ lb per plant. Broadcast the fertilizer evenly around the plant before applying mulch in late winter.

Mulch is very important for growing healthy rabbiteye blueberries. It is required for acidifying and cooling the soil, conserving soil moisture, and controlling weeds. Provide a deep mulch (approximately 6 inches) and extend it at least 2 feet from the crown of the plant. This is extremely important the first 2 years while the plants are establishing.

Various organic materials such as peat moss, pine straw, leaves, and grass clippings can be used. But do not use barnyard manure; it contains toxic salts. If weeds grow through the mulch, remove them by hand or with grass-selective contact herbicides.

Drip Irrigation

You can't grow rabbiteye blueberries successfully without some form of irrigation. No southern fruit crop will profit more from drip-irrigation than rabbiteye blueberries. The volume of water should correspond to season, plant size, and soil texture. Initial spring waterings should be relatively light. Once in full growth, 1-year plants should receive ½ gallon per day. Increase the rate to 1 gallon the second year, adding a gallon per year per

plant to a maximum of 5 gallons per day, or 35 gallons per week. It is better to water once per week rather than daily. Water is especially important during the long fruit-ripening period.

Pruning

Rabbiteye blueberries require little pruning. Lower limbs can be thinned out to keep the fruit from touching the soil, and excessively vigorous upright shoots can be thinned out several feet from the ground to keep the center of the bush open and to keep the bearing surface within reach. Spindly, weak, or dead branches should be thinned out annually during the dormant season.

Harvest

Rabbiteye blueberries are a non-climacteric fruit and should be allowed to ripen on the bush. The fruit of most varieties will ripen over a 4- to 6-week period. A normal season can extend from late May to September. Don't pick the berries until they're fully ripe; otherwise the fruit will be bitter. Once the berries begin to ripen they should be picked every 5 to 7 days—if they aren't, birds will beat you to them. Birds seem to be the key pests with rabbiteye blueberries, and special bird nets are available to protect the fruit.

A mature bush can produce 15 lb of berries (about 9,000 lb per acre).

STRAWBERRIES
Fragaria virginiana

The strawberry can be grown in any home garden in the South. All that's necessary is a little space, full sunlight, and a desire to grow the finest fruit known to man. Strawberries also have very strong pick-your-own market potential.

The strawberry plant is a hard-working little sugar factory capable of producing three times its own weight in fruit every spring. The plant has five basic parts: leaves, crown, roots, fruit, and runners (Figure 28). In early springtime, when days are short and nights are long, the strawberry will produce flowers and fruit. The flowers will bear strawberries in 21 days if a freeze doesn't destroy the fruit. In the summertime, when tempera-

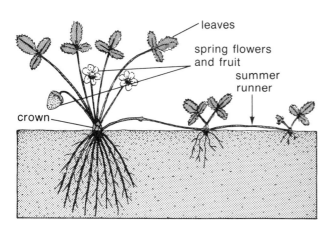

Figure 28. Basic parts of the strawberry plant.

tures are high, days are long, and nights are short, the strawberry plant will send out runners—these are new plants. From June to September one mother plant can send out ten runners with five plants on each runner.

Soil

Strawberries require a well-drained sandy loam soil. If your soil is a silt or clay, you'll have to use a special strawberry soil mix: 1/2 sand, 1/4 peat, and 1/4 soil. The mix can be used anywhere in the South with excellent results.

Your strawberry bed should be at least 6 inches deep and 12 inches wide for each row. Make certain that the bed isn't in a low, flat place and that it drains extremely well.

Irrigation

Strawberries cannot be grown in the South without irrigation. Since the roots are very shallow, they will require frequent watering throughout the year. Sprinkler irrigation is needed at planting, and flood or drip irrigation is needed for the remainder of the year.

Fertilizer

Apply 1/3 oz of 3-1-2 complete fertilizer to 1 cubic foot of strawberry mix or 1 foot of row before planting. Provide a side dressing of 1/3 oz per plant in late March prior to production.

Planting

Use only top-quality, virus-free, nematode-free, salt-free plants. The cost of plants is small compared to other

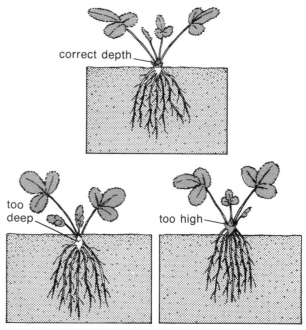

Figure 29. Correct planting depth for strawberries.

expenses involved in successful strawberry culture. Use only a reputable nursery or exclusive strawberry mail-order source for your plants. Runner plants obtained from friends seldom produce as well as certified disease-free plants of the correct variety.

Set the plants in the bed at the same depth at which they were grown, which is mid-point of the crown (Figure 29). When they're set too deep, soil will wash onto the crown and kill the plant. Shallow planting exposes roots, which also kills the plant.

Use a trowel to aid in planting. Don't set plants directly into a spoonful of commercial fertilizer. Push the roots into a "V" hole made by the trowel. If the roots are too long, trim them back with scissors to fit the hole. Press soil firmly around the roots.

Sprinkler irrigate the newly set plants every day for two weeks. Ideally, the plants should be set during cloudy days or late in the afternoon so that they have time to establish before being subjected to hot sun. A good rain right after planting is worth a million. While planting, the roots should not be allowed to dry out. Protect them at all times from exposure to sun and wind or carry them in a pail of water.

Cultural Systems

The South has two basic strawberry climates: a cool winter climate (where the matted row system is used) and a warm winter climate (where the annual system is used).

Strawberries planted in a matted row (left) and the annual system (right). Use matted rows if you get more than 600 hours of chilling; use the annual system if your area is warmer than this.

Matted Row System. This is for areas of the South receiving over 600 hours of winter chilling. These areas are usually too cool to grow sufficient plants in the five winter months.

The strawberry bed should be 42 inches wide and 4 inches high. The plants are set in December, January, or February and will fruit 16 months later as a matted row. Set the plants 18 inches apart. In May, select and "stick" five runner plants from each plant. Try to stick the runner plants in a 20-inch band down the middle of the row. Remove all flowers that develop the first spring: the plant is not strong enough to produce a crop the first spring after planting. To prevent weed growth, mulch the row heavily with pine straw or similar material immediately after planting. The new grass-selective contact herbicides work very well here, because they kill grass weeds without harming the broadleaf strawberries. Strawberries will be produced in April and May of the second year.

Immediately after the harvest season and each year thereafter keep the plants thinned out to approximately 1 foot apart. Try to take out old plants and leave new runner plants each year. The bed will bear for two to four seasons depending on soil, weather, and care.

Varieties for the Matted Row System. Varieties best adapted to the matted row system are Sunrise, Pocahontas, and Cardinal.

Sunrise is a USDA-released variety which produces good crops of medium-sized early berries. The plant is both heat- and drought-tolerant, which is important in the South.

Pocahontas, an old variety from Maryland, has large, firm, tart berries. The plant is resistant to leaf spot diseases and is a good runner producer. It is susceptible to verticillium wilt and red stele.

Cardinal is an Arkansas variety which is very productive and red-stele-resistant. The fruit are medium-sized, firm, and have excellent dessert quality.

Alstar is a new variety which has produced excellent commercial crops in the upper South.

Everbearing varieties were bred and selected in the North and are not well adapted to the South.

Annual System. This system is for the warmer South which, ideally, receives over 290 frost-free days. The success of the annual system depends entirely on how much growth the plants can make during the winter months. If the winter is warm, production will be high; if the winter is cool with late frosts, production can be relatively low. Two types of plants are set, winter plants and summer plants.

Winter plants are typically grown locally and dug in October when the runners stop forming. Winter plants can also be purchased from northern nurseries or from California nurseries, where they are grown in high-elevation, cool-climate nurseries. They are planted the first week in November and are harvested in February, March, and April.

Summer plants are grown in California, dug in February, refrigerated at 28°F until August, and are planted in late August and early September and harvested in March and April.

With either planting date (summer or winter), the plants should be mulched with black polyethylene film 1½ mm thick and irrigated with drip irrigation. The black plastic, irrigation hose, and fertilizer should be placed on well prepared 42-inch beds at least 2 weeks before planting.

Set the plants 12 inches apart on double rows down the middle of the bed. Sprinkler irrigate the strawberry bed every day for 2 weeks following planting.

The refrigerated summer-set plants will send out runners immediately after planting. These must be pruned off as they develop.

These two annual systems have produced as many as 2,000 12-pint flats per acre in the South.

Varieties for the Annual System. Varieties best adapted to the annual system are Sequoia, Chandler, Fresno, Tioga, Tangi, and Douglas.

Sequoia strawberries. They have a marvelous taste but are too soft for commercial production. With these, you can do better than the commercial growers!

Sequoia is a very large, bright red California strawberry which is too soft for commercial production. It is probably the best home variety for the warmer South.

Douglas is a new California variety that has produced heavy yields of very large fruit along the Gulf Coast.

Chandler is an excellent California variety with large fruit, heavy yield, and very good taste for the area south of the 600-hour chilling zone.

Fresno is a large, red, firm California strawberry which produces well as a refrigerated summer plant. It does not have good leaf-spot disease resistance.

Tioga is a large rough, red, firm California strawberry which produces well when planted in either summer or winter. It doesn't have leaf spot resistance.

Tangi is a large, red, firm Louisiana strawberry which is extremely well adapted to winter planting in November. The plants have excellent leaf spot resistance.

Strawberry Pests

Strawberries have many pests and must be watched closely. There is no finer cure for strawberry problems than a watchful eye and early treatment.

Leaf spots are a major problem. Several different fungi can cause them. They are especially serious during wet seasons and on the Sequoia, Fresno, and Tioga varieties. Leaf spots can be prevented with labeled strawberry fungicide sprays applied as soon as the spots begin to appear.

Mites are serious problems on strawberries and will infest every planting if uncontrolled. Daily observations are essential for good mite control. As soon as one plant becomes infected, the entire bed should be treated with a miticide cleared for strawberries. Several follow-up sprays should be made to clear the pest from the bed. Infected leaves will appear dull. Hundreds of very tiny white or gray specks will appear on the leaves of mite-infested plants.

Nursery problems are very common if strawberry plants are obtained from a friend or relative. Purchase only certified disease-free plants to prevent viruses or nematodes. Homegrown plants in the Southwest can also have undesirably high levels of salt accumulated in the crown. You're asking for trouble when you grow anything but certified plants. Approximately 10 nurseries in the U.S. grow certified strawberry plants, and these are the sources you should use.

Special Strawberry Beds

When space is limited or the soil is not optimum, strawberries can be grown in special beds such as pyramids, barrels, borders, or groundcovers. These beds also make an attractive accent for the landscape.

The soil should be the special mix of 1/2 sand, 1/4 peat, and 1/4 soil. Sterilize the soil to kill all weed seeds and nematodes. The bed should be at least 6 inches deep for good root development, high water-holding capacity, and sufficient internal drainage. A typical strawberry pyramid is shown in Figure 30.

Strawberries are perfect as accents or groundcovers. If your soil isn't sandy enough, a raised bed like this is a simple way to overcome the problem.

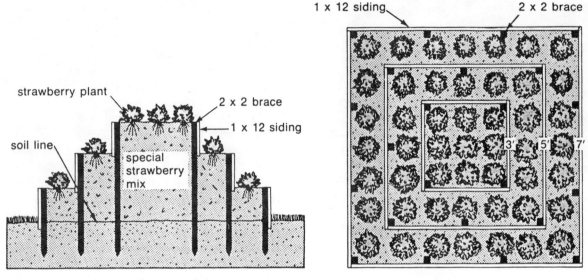

Figure 30. Side and top view of a 49-plant strawberry pyramid.

Harvests

Strawberries will begin to ripen 3 weeks after the last spring freeze. One plant can produce as many as 25 strawberries, or 1 pint. The fruit are non-climacteric and must ripen fully on the plant. Harvests should be made daily if one is to stay ahead of the birds. Bird nets are necessary on small plantings.

If the night temperatures remain cool, southern strawberries can be picked for as long as 6 weeks. However, if late frosts and an early summer combine, harvests may last only 2 weeks. On rare occasions harvests have lasted 10 to 12 weeks.

During dry weather, strawberries should be flood or drip irrigated after every harvest.

Few foods grown in this world are more delicious than the strawberry. They are also extremely easy to process: simply wash and place in the freezer. However, most, if not all, of the berries will be eaten in the field, especially if youngsters are involved in the picking. Pick-your-own strawberry plantings are growing in popularity and are bringing strawberries back to communities that lost strawberry production years ago.

Vines: Grapes, Muscadines & Kiwifruit

GRAPES
Vitis sp.

Interest in viticulture and enology (the growing of grapes and the making of wine) is sweeping the country. Small vineyards are being planted all across the South, and it's no surprise: few crops offer as much challenge and reward as grapes. The South doesn't have the ideal climate for grapes, and this is primarily why southern viticulture hasn't become a commercial industry, except in northern Arkansas, Virginia, Texas, and, most recently, Florida, Tennessee, and New Mexico.

Types of Grapes

European Grapes (*Vitis vinifera*) are the traditional grapes of the Old World, such as Thompson Seedless and Cabernet Sauvignon. The vine is well adapted to mild, hot, dry climates such as California. Vinifera varieties can bear as much as 20 tons of grapes per acre. Vinifera varieties do not have a true rest period and consequently are very susceptible to freeze injury. Massive, rapid-moving cold fronts that follow exceptionally warm periods cause major freeze damage to Vinifera varieties all across the South. The quality of Vinifera table grapes, raisin grapes, and wine grapes is the standard by which other types are compared.

American Grapes (*Vitis labrusca*) are the typical Concord-type varieties grown in the cooler climates of the central and northeastern states. Labrusca vines can bear as much as 8 tons per acre. The Concord variety has become the standard by which grape juices are judged. American grape varieties have a much higher degree of disease resistance and can be grown in more humid areas than the Vinifera grapes.

French X American Hybrid Grapes are the result of cross-breeding European grapes with American grapes, the objective being to obtain the quality and yield of the Vinifera grapes and the hardiness and disease resistance of the American grapes. This has been accomplished to a certain degree with several outstanding hybrids that produce as much as 10 tons per acre of relatively high-quality disease-resistant grapes.

Muscadines (*Vitis rotundifolia*) are the grapes of the South. The vines are disease-resistant and are extremely well adapted to southern climatic conditions. Yields as high as 6 tons per acre have been obtained. The grapes can be eaten fresh, processed as jelly, or fermented into wine. Muscadines are not traditional bunch grapes and are discussed separately in this chapter. Also see pages 84–85 for a discussion of wild grapes and muscadines.

Soil and Climate

Grapes can be grown in many soils, but they grow best in well-drained, deep, sandy loam soil. Shallow heavy clay soil will not produce the vine vigor, tonnage, or quality of better-drained soils.

Grapes do not have the well-defined rest period apples and peaches do. Vinifera grapes will grow any time temperatures and moisture are adequate. For this reason the Vinifera grape is highly subject to winter injury from early fall freezes, midwinter deep freezes, and late spring frosts. American and Hybrid varieties are more cold hardy than the Vinifera varieties. All grapes should be preconditioned to winter freezes by slowing growth in August.

High humidity is a serious grape cultural problem. It stimulates the onset of diseases, particularly black rot and downy mildew.

Vinifera Varieties

Vinifera varieties are only adapted and recommended for the drier and cooler climates west of the 30-inch rainfall line or north of the 1,000- to 1,400-hour chilling zone. The 0°F annual low-temperature line is the upper limit for commercial production of Vinifera varieties. Home plantings can be grown with special attention to freeze protection in midwinter. High humidity, black rot, Pierce's Disease, and freeze injury limit these varieties to this area. The following varieties are all heat-tolerant.

Thompson Seedless is a small, white, seedless table grape which forms large, long clusters. The vine is very vigorous and very high-yielding. It must be cane pruned because it doesn't bear on the first three nodes.

Chenin Blanc is a medium-sized white wine grape which forms large, compact clusters. The vines are very vigorous and productive. It responds well to cordon pruning.

Emerald Riesling is a medium-sized white wine grape which bears as large clusters. The variety has excellent sugar, acid, and pH levels under southern growing conditions. The vines are vigorous and productive.

Cabernet Sauvignon is the premium red wine variety of the world. It produces very low yields of small clusters on a vigorous vine. It can be cane or cordon pruned.

Chardonnay is the premium white wine grape of the world, and it has done surprisingly well in central and West Texas. It can be cane or cordon pruned.

White Reisling is a very cold-hardy and productive grape which makes excellent sweet wines. It can be cane or cordon pruned.

American Varieties

Concord is a large, black table and juice grape which forms medium-sized clusters. The vine is moderately vigorous. Concord can only be grown in the far-northern area of the South receiving over 1,200 hours of chilling. It will produce inferior fruit and have extremely uneven ripening when grown further South.

Fredonia is a large, black table, juice, and jelly grape which forms large, compact clusters. The vine is hardy and vigorous. Fredonia can be grown well in 800- to 1,200-hour chilling zones. It is very similar to Concord but doesn't have a severe ripening problem.

Golden Muscat is a large, white table and jelly grape which forms large, loose clusters. The vine is moderately vigorous and very productive.

Champanel is a large, black jelly grape which forms medium-sized loose clusters. The vine is very vigorous, heat- and drought-tolerant, and resistant to Pierce's Disease (page 60). It is highly susceptible to grape-leaf folder in late summer. Champanel is an excellent grape arbor variety. It can be grown in the 400-hour chilling zone and along the Gulf Coast.

Lenoir (Black Spanish) is a small, black grape, used for wine, juice, and jelly, which forms large, compact clusters. The vine is vigorous, produces moderate crops, and is susceptible to black rot. It can be grown in the warm 400-hour chilling zone. Lenoir vineyards have survived for 30 years in areas where Pierce's Disease destroyed all other varieties.

The Lenoir (Black Spanish) grape seems less susceptible to Pierce's Disease than most other varieties.

Lake Emerald is a seeded, greenish-white Florida jelly and juice variety that is an excellent grape for the sandy soils along the Gulf Coast because of its resistance to Pierce's Disease. It produces medium-sized berries on large clusters.

Favorite is a Black Spanish seedling that is of a better quality than its parent and is Pierce's Disease tolerant. It has dark red juice that is very good for juice, jelly, or wine.

French X American Hybrid Varieties

These hybrids are excellent varieties for areas receiving 800 or more hours of chilling. Pierce's Disease will prevent extended vine life in the warmer areas of the South. There are hundreds of hybrid varieties that have been grown and evaluated in the South. The following are those that grow and produce the best:

Seibel 9110 (Verdelet) is a medium-sized yellow grape for table and jelly. It forms medium-sized, compact clusters. The vines are productive, vigorous, cold-hardy, relatively disease-resistant, and very productive in the upper South.

Seyve-Villard 12-375 (Villard blanc) is a medium-sized white grape which forms large, loose clusters. It is used as a table grape or for wine and jelly. The vine is moderately vigorous, relatively disease-resistant, and very productive in the upper South.

Aurelia is an excellent, very large table grape which forms medium-sized loose clusters. The vine is very vigorous and productive. It produces high-quality seeded table grapes in the upper South.

Seibel 7053 is a medium-sized black wine grape which forms large, compact clusters. The vine is vigorous and productive.

Venus is an early-ripening blue seedless table grape developed by Dr. Jim Moore of the University of Arkansas. It needs to be grown on the deep, sandy soil of the upper South. It has not produced well south of the 1,000-hour chilling line.

Reliance is a new seedless table grape from the Arkansas breeding program. It produces moderate crops of extremely sweet fruit.

Saturn is a new seedless table grape from the Arkansas program. It has great promise for success in the upper South. The red fruit has a good, firm, crunchy texture and excellent taste.

Orlando Seedless is a new seedless white table variety from Florida. It produces high yields of very long, thin clusters. It is resistant to Pierce's Disease and should be the leading table variety south of the 800-hour chilling line.

Blanc du Bois is the first high-quality, Pierce's-resistant wine grape for the deep South. This new white variety was developed by Dr. John Mortensen at the University of Florida. It is vigorous, productive, and makes a fine muscat-type wine.

Rootstocks

Grafted grape vines are protected against nematodes and are adaptable to clay soils. Improved varieties are whip-grafted onto 1-year-old rootstock cuttings in February and grown in the nursery for 1 year before planting in the vineyard (Figure 31).

Dogridge, a seedling selection of *Vitis champini*, is relatively resistant to soil-borne nematodes. The cuttings

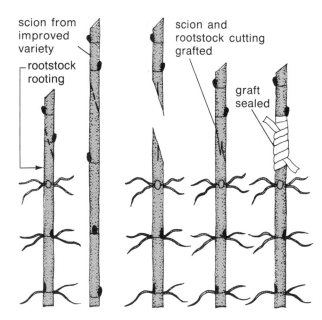

Figure 31. Whip-and-tongue grafting technique for grapes. If the cutting on the left was from an improved variety (instead of a rootstock), it would simply be calloused and planted in the nursery row without grafting.

are relatively easy to root and are compatible with most improved varieties. The vines are extremely vigorous and frequently have serious sucker problems.

Champanel is a T.V. Munson selected hybrid of *Vitis champini* X *Vitis labrusca* var. Worden, which is vigorous and well adapted to heavy clay soils. The cuttings root extremely well and are compatible with most improved varieties. Champanel is somewhat tolerant to cotton root rot on high-pH calcareous soils.

Tampa is a nematode-resistant rootstock from Florida that is adapted to deep and sandy soils.

SO4 is a German rootstock that reduces vigor and helps improve cold-hardiness of vines north of the 1,000-hour chilling line. It has nematode resistance, and is gaining in popularity in numerous commercial wine-growing areas.

Propagation

Stem cuttings are the major means of propagating grape varieties and rootstocks. Most varieties root very easily compared to other fruit crops. During January or February pruning, collect moderately vigorous, healthy, disease-free, straight 3/8-inch cuttings 12 to 14 inches long. Cut the basal end square 1/2 inch below a node (Figure 31). Make the terminal cut on a slant 1 inch above the upper node. Good cuttings usually contain four nodes. Remove the buds at the three lower nodes to prevent sprouting in the callus trench, nursery, and vineyard.

Group and label the cuttings for callusing in a soil trench before lining out in the nursery row in early April. (For illustrations and details on callusing, see Figures 44–46, pages 76–78.) The cuttings will have callused and begun rooting when removed from the callus trench. The cuttings, once callused, should be planted with only one bud above the ground. The nursery row should be kept weed-free and well watered.

The cuttings should root rapidly and grow a strong plant in one season. If it is an improved variety, transplant to the vineyard in February. If it is a rootstock, it should be whip-grafted and grown for a second year before transplanting (Figure 31).

Planting and First Year's Growth

New vines should be planted 8 × 12 feet in January, February, or March with only one or two buds above the ground (Figure 32). Trim the roots to 2 to 4 inches in

length and dig the hole only slightly larger than the trimmed root system. Let the vine grow untrained the first year. Remove weeds as they appear or prevent them with black plastic mulch. Drip irrigate with 7 gallons of water per week from April 1 to August 15.

Second- and Third-Year Training

Cut the vine back to two buds before the second growing season. More than two shoots will grow very rapidly in the early spring. Select one main shoot and tie it to the stake as it develops (Figure 33). For cane pruning (Figure

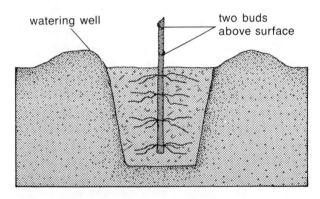

Figure 32. A properly planted grape vine.

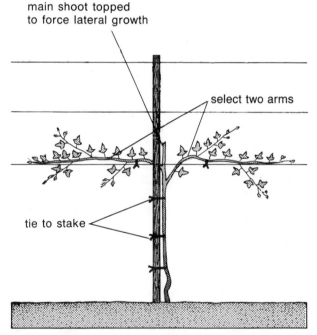

Figure 33. A properly trained two-year-old grape vine.

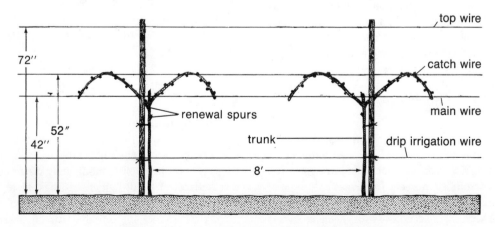

Figure 34. A mature grape vine trained and properly pruned to the 2-cane system. (Vines pruned to 20 buds each.)

34), top the shoot at the top wire; for cordon pruning (Figure 35), top the shoot at the lower wire.

The three wires on the trellis should be 72, 52, and 42 inches or 60, 46 and 36 inches from the ground. The taller trellis is recommended because it exposes more of the vine to sunlight and is easier to prune and harvest. Most trellises also have a wire at 18 inches to attach the drip irrigation line. A farm fence "T" post should be placed at each plant to train and support the vine. Large support posts will also be needed every 12 vines down the row. The ends of the trellis should be well anchored. The trellis wire should be 12$\frac{1}{2}$-gauge smooth galvanized wire.

Before growth begins the third year, prune the young vines back to only 6 to 12 total buds, depending on vine vigor. This will be 2 to 3 buds for each of the 2 canes, or 3 to 6 buds on each arm of the cordon. Only 1 cluster should be allowed to fruit from each shoot. All other clusters should be thinned out in early May.

Thinning out clusters on a grape vine. This is absolutely essential to ensure health, vigor, and regular production.

Pruning Mature Vines

A variety should be pruned according to the plant's vigor and production potential. In general, the vines will tell you how many buds to leave after pruning. Leave the same number of buds as canes produced that season. If in January, before pruning, a vine has 16 canes that are approximately $\frac{3}{8}$ inch in diameter, leave 16 buds. Do not count canes that are less than $\frac{1}{4}$ inch in diameter. If there are canes of $\frac{1}{2}$ inch in diameter, leave 2 buds rather than one. Leaving too many buds is one of the most common errors in grape culture. If 90 percent of the vine is not removed each year, the vine will overbear, become weak, and die.

Cluster Thinning

Overcropping is a serious problem with young vines and must be corrected by cluster thinning to insure healthy vines, quality grapes, and regular production. Three-year-old vines should be thinned to a maximum of 8 to 12 clusters; four-year-old vines should fruit only 16 to 24 clusters. Mature vines should carry 40 to 80 clusters depending on cluster size and vine vigor. The point is to prevent overcropping. This is extremely important if you don't have a good soil, weed-control, irrigation, and disease-control program.

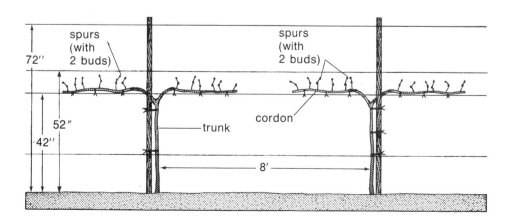

Figure 35. A mature grape vine trained and properly pruned to the bilateral cordon system. (Vines pruned to 25–30 buds each.)

Black rot is a serious grape disease that can be controlled with early fungicide sprays.

Pierce's Disease is another serious grape disease; resistant varieties are the only control.

Cultural Practices

Fertilizer is not a serious limiting factor in the early life of the vineyard. On sandy soils, 1/4 lb of complete 4-1-2 fertilizer can be applied each February. Acid soils will need lime.

Drip irrigation is excellent for grape culture. Apply 7 gallons a week per vine per year of age until you're applying 35 gallons of water per week. The drip system must be cut off in early August to harden off the plants before the first frost. This is especially necessary with Vinifera grapes.

Leaf pruning in mid-May is an excellent tool for improving table grape and wine grape quality. Simply remove the lower 3 or 4 leaves on the shoot. This will allow greater air movement and light penetration. It will also improve the quality of next year's buds, which increases cluster number and size. The most important advantage is that it allows good fungicide coverage for black-rot control.

Grape Pests

Black rot is a serious fungus disease that infects the fruit, leaves, and canes. During high-humidity conditions, spray every 7 to 10 days with Bayleton, Benlate, or Ferbam. Early sprays are much more effective in heading the fungus off before it builds up. Leaf pruning also helps reduce black rot.

Pierce's Disease is the most limiting factor in southern grape culture. It is a very serious disease in areas receiving 700 or less hours of chilling each year, and it can occur in areas receiving 700 to 1,000 hours of chilling, though the probability is less. The disease is caused by a microscopic bacteria-like organism, which is spread by a special group of leafhoppers. The leafhopper carries the disease from common southern grasses, such as Bermuda grass and Johnson grass, to the grape vines.

The best means of control is to plant resistant or tolerant varieties such as Orlando Seedless, Blanc du Bois, Lake Emerald, Lenoir, Champanel, or muscadines. A good insect-control program reduces the probability of disease. Symptoms of Pierce's Disease are very similar to water stress symptoms (see photo). The leaves first show marginal die-back, the canes will develop dead areas, and the vine will go into a gradual decline and die.

Harvest

Grapes are non-climateric fruit and must ripen to full maturity on the vine. Harvests usually begin in mid-July and continue to September. Grapes produce the best wine when their sugar content is 22 to 24 percent and their acid content is 0.7 to 0.8 percent. If the grapes are allowed to remain on the vine too long, sugar content will be too high and acid content will drop too low.

Mature southern grape vines can produce as much as 25 lb per vine or 6 tons per acre. In general, Vinifera varieties are high yielders, hybrid varieties are moderate yielders, and American varieties are low yielders.

MUSCADINES
Vitis rotundifolia

Muscadines are a part of the southern environment and tradition: Southern homemakers, like the early pioneers and settlers, use muscadines fresh and for homemade jelly. The native vines are commonly found in all of the acid, sandy soils across the South, and their berries serve as food for birds and other small animals.

The legendary viticulturist T.V. Munson reported that members of Sir Walter Raleigh's colony found the original Scuppernong muscadine vine on an island in the Scuppernong River in North Carolina in 1554. The muscadine has since become a popular home crop across the South because of natural adaptability, resistance to insects and diseases, and long vine life. Muscadine vines are excellent accent pieces for the landscape, used as trellises, or as a border. The fruit are excellent fresh or as jelly. The vines are very productive and require little care.

Soil and Climate

Muscadines are adapted to fertile, sandy loam soils that are relatively acid. They will grow and produce well in most southern soils, but they perform best in sandy loam soils that are well drained and well aerated. Muscadines will also grow in slightly alkaline soils; however, vine production in these soils is low. These soils will also need frequent applications of iron chelates to prevent iron chlorosis.

Muscadines are a southern crop requiring warm winters—they cannot be grown in areas receiving winter temperatures below 5 to 10°F. This limits them to the areas receiving less than 1,200 hours of chilling. When grown on well-drained soils, they can withstand exceedingly high levels of annual precipitation. Unlike bunch grapes, complications from high relative humidity are not a major limiting factor.

Varieties

Muscadine varieties are selected according to pollination, production, fruit characteristics, and vine vigor.

Fry is a large bronze table muscadine with an outstanding taste. The vines are very productive and disease-resistant. Fry requires a pollinator.

Carlos is a small bronze wine muscadine which forms a medium-sized cluster. The vine is vigorous, relatively disease-resistant, and very productive. Carlos forms a complete flower and does not require a pollinator.

Higgins is a very large reddish-bronze table muscadine which forms large clusters. The vine is very productive, vigorous, and disease-resistant. Higgins is a pistillate (female) variety and requires a pollinator.

Magnolia is a medium-sized bronze muscadine which forms medium-sized clusters. The vines are moderately vigorous and productive. Magnolia forms a complete flower and does not require a pollinator.

Cowart is a very large black muscadine which forms very large clusters. The vine is vigorous and very productive. Cowart forms a complete flower and does not require a pollinator.

Regale is a new red female variety which produces a high yield of top-quality muscadines in large clusters.

Jumbo is a very large black muscadine which forms large clusters. The vines are vigorous and disease-resistant. Jumbo will ripen over a long period of time. Jumbo is a pistillate (female) variety and requires a pollinator.

Jumbo, a black, large-cluster muscadine.

Propagation

Muscadines are propagated by layering. Hardwood cuttings, when separated from the mother vine, will not form roots; shoots must be rooted while attached to the mother vine. Layering is done by taking a lower cane

Propagating a muscadine vine by layering: a shoot still attached to the mother plant is embedded beneath the soil during the dormant season.

A year later, the rooted cane is removed from the soil as a new vine.

and placing it under the soil during the dormant season. After one complete growing season under the soil, this cane will have roots. During the second dormant season, the cane is removed from the soil and a new plant is obtained (see photos).

Spacing and Planting

Muscadines are planted in the dormant season, usually the months of December, January, or February. Space the vines 20 feet apart on 12-foot rows. Dig a hole to receive the entire root system and pack the soil well around the roots. Cut back the top part of the dormant plant to approximately two buds. Start with large healthy plants approximately the size of a pencil. Bring the strongest shoot up onto the trellis the first year and begin training it during the second growing season.

Training and Trellising

Muscadines are trained onto a one-wire trellis 5 feet above the ground during the second growing season. Posts should be set at every 20 feet, with vines between the posts. Use a small wire or stake to train the young vine onto the wire.

Select one vigorous shoot to grow up the stake the second year, and pinch off lateral shoots as they develop. This pinching facilitates the development of one strong, upright shoot, which will develop into the permanent trunk of the muscadine vine. Allow this main shoot to grow upward until it reaches the top wire. Once it is 5 feet, cut this shoot back to force lateral shoot develop-

ment. Then select two lateral cordons 6 inches or more below the wire and train them down the wire. In some instances the cordon can develop to the full length of the wire during the second growing season.

Pruning

Begin pruning muscadines during the third dormant season after the training is complete. Select short 1-year-old spurs along each cordon. The spurs should be approximately 6 inches apart and each spur should be pruned to 3 buds. Figure 36 shows the proper pruning procedure for mature muscadine vines.

Arbors

Muscadines make beautiful arbors. (A typical 12 x 12-foot design is illustrated in Figure 37.) Train the vine up the arbor post the first 2 years. The third year, establish a cordon down cross-members spaced 24 inches apart. Prune the cordon to 3 bud spurs every 6 inches in January or February. The cordons should not be closer than 48 inches. If you're only using one variety, make certain that it is a self-fruitful variety and does not require a pollinator.

Harvest

Muscadines are harvested when fully ripe—from late August to October. Simply remove each berry by hand as it ripens. If you have a larger planting, you can place a

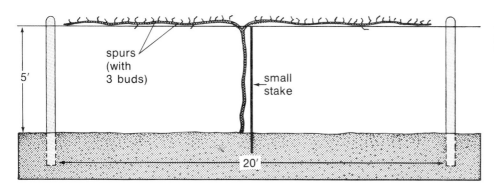

Figure 36. A properly pruned mature muscadine vine.

spurs (with 3 buds)

small stake

5'

20'

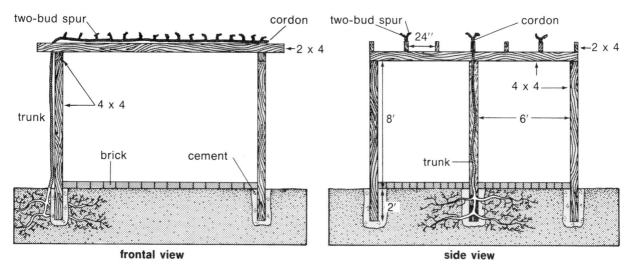

two-bud spur

cordon

2 x 4

4 x 4

trunk

brick

cement

frontal view

two-bud spur

cordon

24"

2 x 4

4 x 4

8'

6'

trunk

2'

side view

Figure 37. A 12 × 12-foot grape or muscadine arbor. This attractive structure will provide welcome shade on hot summer days and edible fruit as well.

catching frame under the vine and bump the arms. This causes the ripe fruit to fall onto the catching frame. You'll need to do this two or three times before you get them all. Birds and wildlife love muscadines, so be prepared to get your share.

KIWIFRUIT
Actinidia chinensis

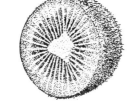

This beautiful, emerald green salad and dessert fruit has taken America by storm. In the last ten years it has become a major commercial crop in California. During the winter months the fruit is imported from New Zealand, where it has been popularized and made into a major fruit crop worldwide.

Unfortunately, the kiwifruit has very exacting climactic and cultural requirements and it is going to be ex-

tremely difficult to grow in the South. Arlie Powell has had good success for several years with kiwifruit in the Mobile, Alabama area. The vine has a chilling requirement of 500 to 700 hours, which limits its culture south of central Florida and in the lower Rio Grande Valley of Texas. At our present state of the art, it appears the kiwifruit will be limited to the Atlantic shore and Gulf Coast, and to no more than 150 miles inland. It is very susceptible to freeze injury if attempted further north. The vines are very heat, drought, and salt sensitive, which creates a real challenge for those brave enough to attempt them. The leaves are large and fleshy and transpire a great volume of water each day; consequently, only the best, deep, well-drained soils and salt-free water will be able to support kiwifruit growth.

The major variety is *Hayward*, which needs a male pollinator for every six Hayward vines. The vines are trained up onto a horizontal trellis at least 5 feet above the ground. Each plant has a single trunk, which is trained into two horizontal trunks, or cordons, down the row at the top of the trellis. The fruiting canes are

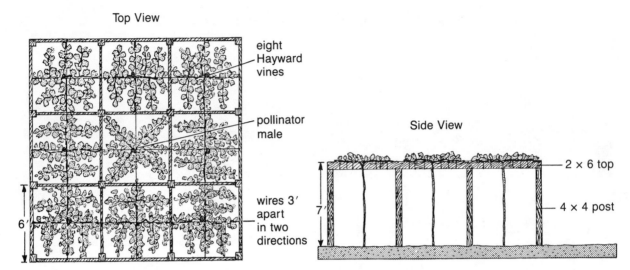

Top View

eight
Hayward
vines

pollinator
male

wires 3′
apart
in two
directions

6′

Side View

2 × 6 top

4 × 4 post

7′

Figure 38. A home garden, 9-vine, kiwifruit pongola.

produced from these cordons, which is very similar to bilateral cordon training in grapes, pages 58–59. Each year the cordons are extended 2 or 3 feet until they meet the cordon of the next vine. The vines should be 18 feet apart in rows that are 20 to 24 feet apart. The cordons and fruiting canes are supported by at least five 12.5 gauge, high-tensile, galvanized wires. Two "T" top or "H" top trellis braces will be needed for each vine. For home plantings, a 9-vine arbor or *pongola* can be constructed in an 18-foot square, with three female Hay-ward vines on each side and a male pollinator in the middle. (See Figure 38.) The pongola should be 7 feet above the ground to allow easy passage.

Since very few kiwifruit have been produced in the South, it is too early to speculate on the production levels.

The hardy kiwifruit, *Actinidia arguta*, has been promoted as cold tolerant for the North; however, it has had severe heat and drought-stress problems when grown in the South.

Nuts: Pecans, Black Walnuts, English Walnuts, Almonds, Chestnuts & Pistachios

PECANS
Carya illinoensis

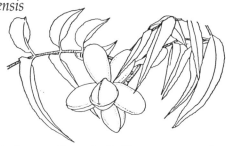

Pecans are truly a southern fruit. They were a major food for Indians thousands of years before Europeans settled in America. Large native trees grow wild along rivers and streams in Tennessee, Arkansas, Oklahoma, Mississippi, Louisiana, and Texas (see page 83). Large pecan-development companies, using primarily the Stuart variety, were responsible for starting the large commercial pecan industry in Georgia, which is outside its native range.

Pecans not only serve as an excellent fruit crop; they are equally useful as shade trees and landscape specimens. Pecan trees grace many homes and landscapes across the South.

Climate

The pecan grows extremely well all over the South. The tree can be grown in climates receiving 10 to 60 inches of rainfall annually. The winter chilling requirement is approximately 500 hours; however, this is not obligatory, as the tree can survive equally well with 300 hours or 1,000 hours of chilling. In the more humid region along the Gulf Coast the pecan requires special attention to prevent pecan scab disease. The pecan grows extremely well in the lower South, where the growing season is over 250 days.

Soil

Good pecan soil is deep, well drained, and fertile. This does not exclude poorer soils; it only limits *commercial* pecan production to the very finest soils. Homeowners can grow beautiful trees for 50 years on soils less than 32-inches deep, but optimum management will be required if the trees are to bear regular crops. The pecan can't stand wet feet and shouldn't be planted in flat, low places that don't drain well.

Varieties

Select your varieties carefully—you'll have to live with them a lifetime. Disease resistance, fruit production, fruit quality, tree strength, and attractiveness are very important considerations.

Non-grafted seedlings make excellent yard trees. However, they will be very slow to bear (10 to 20 years), and the quality of 99 percent of the trees will be inferior to that of grafted trees of improved varieties. Seedling trees develop a strong central leader with no training.

Trees of improved varieties are obtained by grafting seedling trees, or by purchasing trees which were grafted at the nursery. Old varieties can also be topworked by grafting to a new variety.

Pecans are broadly grouped as eastern or western varieties by their resistance to pecan scab disease, with the eastern varieties being very resistant. Eastern varieties can be grown in the East and the West, but western varieties can be grown only west of central Texas.

Choctaw has excellent disease resistance and fruit production, and it is a strong tree. The foliage is large, attractive, and dark green. The Choctaw kernel is extremely high in quality: it is large, light-colored, and delicious. This pecan requires approximately 7 years to come into production.

The excellent nut of the Choctaw pecan. The tree, too, is excellent—it is large, strong, and has attractive dark green foliage.

Cheyenne has good disease resistance and high production. The tree is small, with long, narrow limbs. The kernel is very high-quality, with medium size and bright cream color. Aphids feed on Cheyenne more than on any other variety. The Cheyenne requires 5 years to bear nuts.

Sioux is a very small western pecan with excellent kernel quality. The foliage and shucks have fair disease resistance and the trees will produce moderate crops. The Sioux variety is a very strong tree that is easy to train. It will begin to bear pecans in the sixth year.

Cape Fear is a North Carolina variety that has good disease resistance. It produces high yields of pecans of average size and kernel quality at an early age. The tree is very vigorous and easy to train. Cape Fear is susceptible to foliage scorch.

Caddo is a very small pecan of high quality and excellent disease resistance. It produces pollen very early for protandrous varieties. The tree is strong and comes into production at an early age. Caddo ripens in late September and can be grown above the 1,200-hour chilling line.

Wichita is the most productive, high-quality pecan in the world. It can only be grown in the West because it is highly susceptible to pecan scab. It is almost as difficult as a yard tree because of its high demand for soil, space, water, zinc, and nitrogen. The tree is very difficult to train and is subject to freeze injury because of its very low chilling requirement. Heavy limb breakage is common for Wichita trees.

Pawnee is a new USDA variety developed by Dr. Tommy Thompson, which shows great promise because of its early ripening, aphid resistance, production, and quality.

Maramec is a new Oklahoma variety adapted to the upper limits of the South. It is cold tolerant and produces moderate crops of fair-quality pecans. The tree is strong and attractive.

Melrose is a new Louisiana variety that produces a strong tree. The nut is of average size and has fair-quality kernels.

Desirable is a very old variety noted for regular production and large, high-quality kernels. The tree has fair disease resistance and produces a crop every year. Desirable foliage is light green, and the limbs are only moderately strong.

Elliot is an old variety from Florida that has extremely small pecans. Excellent disease resistance is its outstanding trait. The tree is very slow to come into production, but it is strong. Elliot is recommended as a lawn tree only for highly humid areas and where the trees cannot be sprayed.

Western is the leading variety for the far west. It is highly susceptible to pecan scab and shouldn't be grown east of central Texas.

Varieties to Avoid. The *Stuart, Mahan,* and *Burkett* varieties are not recommended for planting because newer varieties are superior. With good management, mature trees of these varieties perform well.

The *Success* variety should *not* be planted. It has a physiological disorder called shuck disease, which is a very serious, uncontrollable problem.

Variety Pollination. Flower buds on pecans are the result of good, healthy leaves in late August, September, and October. These leaves manufacture food, which is used to initiate flowers the following spring. Flowers on pecans are monoecious (separate male and female flowers on the same tree). Some varieties develop male catkins first and are called *protandrous*. Other varieties

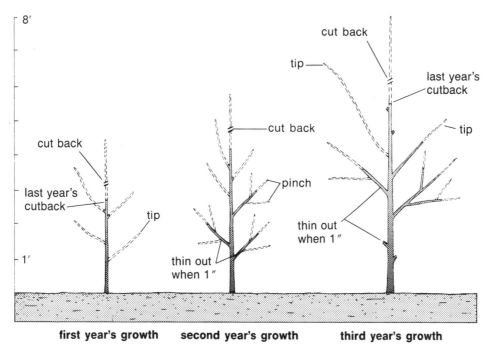

Figure 39. How to train pecans after each year's growth.

Table 8
Pecan Variety Pollination Types

Protandrous Type (pollen first)	Protogynous Type (nutlet first)
Cheyenne	Choctaw
Desirable	Sioux
Western	Elliot
Cape Fear	Wichita
Caddo	Maramec
Pawnee	Melrose

produce the female nutlets first and are called *protogynous*. Each type cross-pollinates the other naturally. Table 8 presents the varieties of each type. If you're planting two pecan trees, plant one of each pollination type.

Spacing

Pecans mature into very large trees and require space. The root system of a mature tree extends at least twice the distance of the limbs. If the limbs of two trees are touching or crowding, the roots have long before been crowded.

Pecans should not be planted closer than 35 x 35 feet. A practical spacing for pecan trees is 50 feet apart. If you plant two or more trees, a temporary or filler tree can be planted in the middle. When the trees begin to crowd, the temporary tree should be removed in 15 to 20 years, leaving the permanent spacing of 50 x 50. Tree removal is the best means of thinning pecans. Do not dehorn the trees by severe pruning (Figure 53).

Training and Pruning

Train your young pecan tree to form a central leader tree (Figure 39). This will make the tree strong and allow maximum utilization of space.

To train a central leader, cut back one half of the long stem at planting. This will force growth into several central leaders and also keep the stem in balance with the roots. If the top is not cut back severely, growth will be weak and the tree will transpire more water than the roots can absorb.

During the first three growing seasons, growth should be forced into the central leader by removing all but one leader in May and pinching the lateral shoots as they develop. Pinching is necessary at least once in June, July, and August. Simply pinch out the soft, green, growing tip on lateral shoots 18 to 24 inches long. During each dormant season, cut back the central leader by one half. This will continue to force several leaders near the cutback point. In May or early June, select the most vigorous shoot of the group to become the central leader and thin out the outer shoots near the cutback point. Do not leave more than one central leader. If you fail to thin out these extra leaders in the summer, it can be done the following February. Thin out the lower lateral side shoots only when they reach 1 inch in diameter, not before. These side limbs develop what is called a "trashy trunk," and they are essential in stimulating strong root growth and rapid central leader development.

Leave permanent scaffold limbs approximately 5 feet from the soil line and do not pinch-prune them. After the fourth season, tip-prune all scaffold limbs which make over 24 inches of growth. You tip-prune by removing less than 2 inches of the very tip of a shoot in the winter months. All narrow-angled crotches and crow's feet should also be thinned out during the winter months (Figures 40 and 52).

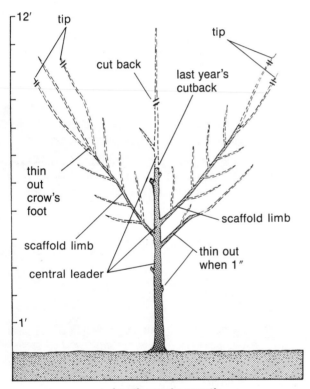

fourth year's growth

Figure 40. Tip-pruning the pecan tree for central leader development.

Fertilization

Young Trees. Pecans respond extremely well to frequent, small amounts of ammonium sulfate or ammonium nitrate fertilizer. It can be applied the first year if the trees are growing fast. Spread 1/2 lb of nitrogen fertilizer around the tree in June. The second year apply 1 lb in April, May, and June. The third year, apply 2 lb in April,

May, and June. From the fourth to the seventh year, apply 4 lb per tree. Always spread the fertilizer away from the trunk, past the drip line in the root absorption zone.

Mature Trees. Once the trees begin to bear over 15 lb of pecans per tree, or after the trees are 6 inches in diameter, apply 1 lb of nitrogen fertilizer for each inch of trunk diameter in late March and again in May. After the tree is in full production, all non-fruiting shoots should make no more than 8 to 15 inches of terminal growth each year. If more growth is made, reduce the fertilizer rate.

Never fertilize after the month of June because it can stimulate late-season growth and encourage freeze injury. This is especially true for young trees under drip irrigation.

Zinc

All pecan trees will respond to foliar zinc sprays of powdered zinc sulfate or liquid zinc nitrate. Your pecan trees need zinc for shoot elongation and leaf expansion. Zinc sprays are absolutely essential when the soil pH is above 7.0. Zinc chelates or zinc injectors are not effective and should not be used.

Large pecan trees in Louisiana, Arkansas, Oklahoma, and Texas require at least three zinc sprays each spring. Trees in central and west Texas require five foliar zinc sprays for optimum growth and production. Small trees less than 6 years of age should be sprayed with zinc every 2 weeks from April 15 to August 1 each year.

Do not spray zinc onto any other fruit crop, for it can seriously damage the foliage. Never let zinc spray drift onto peach trees—it will kill them.

Propagation

Pecans do not grow true to type when grown from a seed. Grafting is required to propagate an improved variety. Varieties are commonly propagated in nurseries by patch-budding improved varieties onto 2-year-old seedling rootstocks in August and September. The variety top is grown for one season. The tree is then dug and sold with a 1-year top and 3-year root. In the Southeast, pecan nurserymen usually whip-graft 2-year-old seedlings during the winter months. This cannot be practiced in the more arid climates of the Southwest because the grafts dry out before growth begins.

Pecan growers, both home and commercial, frequently plant seedling trees and graft them 3 or 4 years later. Homeowners frequently plant pecan seeds or have volunteer trees come up in the yard. These trees can also be grafted.

The influence of zinc sprays on pecan leaf size: no supplemental zinc (left); with zinc sprays (right).

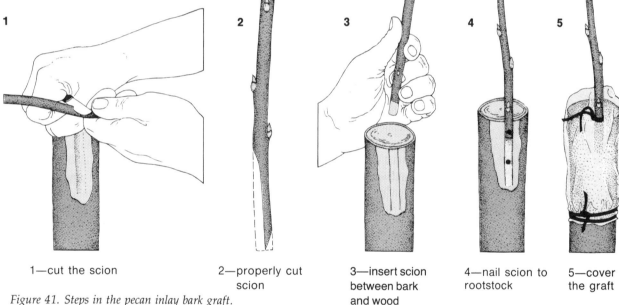

1—cut the scion

2—properly cut scion

3—insert scion between bark and wood

4—nail scion to rootstock

5—cover the graft

Figure 41. Steps in the pecan inlay bark graft.

To graft, you must collect and store graftwood in January or February before the variety trees begin to grow. Store it in very slightly moistened paper towels sealed in polyethylene bags at approximately 45°F until grafting in the spring. You will need to graft pecans while the bark is slipping in April and May. Place the graft on the windward side of the rootstock so the wind will blow the graft into the tree.

Bluefford G. Hancock is recognized for having promoted the Texas pecan inlay bark graft now used successfully by thousands in grafting pecans. The scion, or graft stick, should be cut so that it will fit flat against a 2-inch rootstock limb (Figure 41). Insert the scion between the bark and wood of the rootstock, as shown. There should be absolutely no air space between the flat cut of the scion and the wood of the rootstock. Secure the scion to the rootstock with short wire nails, 9/16-inch wire staples, or flagging tape. Cover the graft first with aluminum foil and then with a polyethylene bag to reflect light and retain moisture. Make drainage cuts on the lower edge of the polyethylene bag to prevent water from trapping in the bag. Remove the tape, poly bag, and foil 6 weeks after the graft has taken and begun to grow.

If excessive growth occurs the first or second year, wind can blow the grafts out of the tree. To avoid this, prune the new graft shoots to approximately 24 inches or tie them to stakes.

Pecan Pests

Frequently check your pecan foliage and nutlets to determine if an insecticide or fungicide spray is needed. Your county Extension agent can inform you of the major pests and when to check for them. Zinc sprays should be applied, even if no insecticide or fungicide is needed.

Pecan scab is by far the most prevalent pecan disease in the South. It is extremely severe in areas of high rainfall (such as along the Gulf Coast). The fungus develops as small black spots on the leaves and shucks in early spring. If unsprayed, it can infest the entire shuck, turning it completely black. Pecan scab is easiest to prevent early in the season. Sprays late in the year are not nearly as effective as early sprays. When very dry conditions prevail, scab growth usually stops; however, it will begin again with the rains.

Stick Tights is a relatively new pecan disorder which affects the pecan when water begins to fill the nut in late July or early August. The fungus turns part of the shuck black and causes the nuts to drop or be poorly filled. Stick tights can occur in both wet and dry climates. It can be reduced with benomyl sprays during and following the waterstage.

Pecan nut casebearer is a small insect which feeds on the nut and causes it to fall from the tree prematurely. Its damage is easily recognized because it enters the pecan only on the stem end of the small fruit. A generation usually appears in early May and a second generation occurs 42 days later. Proper timing of the casebearer

Salt burn of pecan foliage. This happens when the trees are irrigated with lawn sprinklers and the water supply has an appreciable amount of dissolved salts. Avoid wetting the leaves; use a hose or drip irrigation and wet the soil.

spray is important. Contact your county Extension agent for the exact date to spray in your area.

Pecan weevil is a very serious insect pest of pecans in the South. Fortunately, it is only located in certain areas. The insect emerges from the soil in mid- to late August after a heavy rain and lays an egg in the pecan when the kernel is dough-like. At harvest, the egg has hatched into a short, fat, white grub with a red head. The grub will eat all of the kernels. Then it will eat a hole about 1/8 inch in diameter in the shell and return to the soil. Carbamate insecticides labeled specifically for the pecan weevil will control the pest.

Aphids, very small insects which feed on pecan (and many other) leaves, are a real nuisance to growers and homeowners. They can be recognized easily by a sticky honey dew that accumulates on the surface of the leaves and the ground under the trees. The aphid can cause very serious damage to the foliage from early June to September. They can be controlled with a labeled insecticide when they appear. The overuse of insecticides has created a major problem with aphids on pecans. The best long-term solution is very limited use of insecticides, and biological control using lady beetles and lacewings.

Fall webworms produce a thick web or tent in the trees in August, September, and October. The insect can literally eat every leaf on a tree if uncontrolled. Prune out the first webs as they appear. If the infestation becomes extensive, a labeled insecticide spray will be needed.

Salt burn is a major problem on trees in the South that are irrigated with lawn sprinklers. It is particularly acute when the water contains more than 500 ppm soluble salts. The problem is evidenced by marginal and tip leaf browning. Salt burn can be corrected by using drip irri-

Pecan scab. This disease is extremely severe in high rainfall areas such as the Gulf Coast.

A healthy 15-year-old pecan tree can produce 100 lb of nuts a year. Long poles are a practical way to get the nuts to the ground.

gation or simply soaking the soil with very slow running water rather than using a sprinkler.

Harvest

Pecans begin to ripen in mid-September until December. They require a minimum of 200 frost-free days to properly mature a crop. In more northern areas, small short-season pecan varieties must be grown. As the shucks begin to open, the nuts can be thrashed to the ground with long poles or mechanical tree trunk shakers.

Pecans vary in size, from 30 to 90 nuts per pound. The nuts can range from 40 to 60 percent kernel. A healthy 15-year-old pecan tree on good soil can produce 100 lb of nuts annually. Pecans can be stored in-shell for 6 months in your refrigerator and indefinitely in your deep-freeze.

will make an extremely attractive, stately, home landmark. Its timber is also an important economic crop in Arkansas, Kansas, Oklahoma, and Missouri.

Black walnut trees grown from seed will require 10 to 20 years to bear, while grafted trees bear in 5 to 10 years. *Thomas* is the most widely planted grafted variety in the South because it has higher yields and nuts with thinner shells and higher quality.

Harvest the nuts as soon as they fall from the tree. Peel the shuck off, wash off the remaining shuck, and air dry in the sun in onion sacks for 3 weeks.

ENGLISH WALNUTS
Juglans regia

BLACK WALNUTS
Juglans nigra

The black walnut is native to all southern states except Florida. In addition to producing its tasty nut, the tree

Oddly enough, English walnuts have come to us by way of Poland, Russia, Czechoslovakia, Germany, and France. They are quite frequently called Persian or Carpathian walnuts. The trees are very vigorous and grow into attractive yard specimens with a distinctive silver-shaded bark.

Unfortunately, few varieties have produced as well in the South as they have in California and Europe because of a bacterial disease called walnut blight, which infests the young nutlets during and immediately following the bloom stage. Loy Shreve, of Texas A&M University, has traveled to China, Poland, Rumania, and Hungary to collect genetic material which appears to be resistant to walnut blight. Early results are very promising and several new varieties are now being propagated for trial throughout the South. The best varieties to date are *Reda, Sibisel Precose, Miles, S2,* and *Sejveno.* He has also determined that the native Texas Black Walnut, *Juglans microcarpa,* is a far superior rootstock for walnuts growing in the high-pH soils of the arid Southwest. In the acid soils of the Southeast, the Eastern Black Walnut, *Juglans nigra,* continues to be the best rootstock.

During periods of prolonged humidity in the spring and summer, walnut blight damage can be reduced with Kocide sprays. Begin when the catkins are approximately 1-inch long and when the humidity is high.

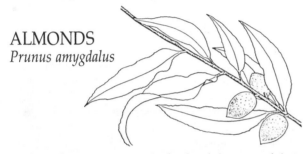

ALMONDS
Prunus amygdalus

Almonds are not grown in the South because of their very specific environmental requirements. The trees bloom extremely early, frequently loosing the entire crop to freezes. The fruit require an extremely long, dry summer, which we seldom receive in the South. As if this were not enough, almonds are also susceptible to brown rot and bacterial stem canker, and they have cross-pollination problems.

There is some interest in several newer seedlings in the arid Southwest.

CHESTNUTS
Castanea sp. hybrids

When the American chestnut was killed by chestnut blight between 1910 and 1940, many thought we would never see chestnuts in the South again. At the turn of the century, the USDA made several Chinese chestnut plantings across the nation. Following the work of at least two generations of researchers, we are for the first time seeing a new birth of chestnuts in the South. There are still major problems with fruit quality and resistance to the chestnut gall wasp; however, many southerners are planting the new varieties and looking forward to even better varieties. Hubert Harris, of Auburn University, has released three varieties from his original USDA planting. These include *Cropper, Leader,* and *Homestead.* Robert Dunstan Wallace, of Alachua, Florida, has released the *Revival* chestnut. All of these appear to be resistant to chestnut blight and produce very edible fruit.

These new varieties need to be tested in every area of the South. As with other large trees, they need to be planted at least 30 × 30 feet apart to prevent overcrowding as the trees age. Deep, well-drained soil will grow the strongest tree.

As with other new fruit crops in the South, only time will tell if the chestnut can again become a major fruit crop.

PISTACHIOS
Pistachia vera

The pistachio originated in the Middle East, but has gained much popularity as a new crop in southern California and Arizona. New trial plantings are being evaluated in Texas west of Fort Stockton and the 10-inch rainfall line. The pistachio is a true desert crop and cannot tolerate any rainfall in August and September. The major variety being planted is the *Kerman,* which is grafted onto a *P. terebinthus* seedling rootstock. The trees are spaced at least 20 × 20 feet. They are considered drought tolerant, but some irrigation will be needed in the growing season. The fruit are harvested in September.

Subtropical Fruits: Citrus, Figs, Persimmons & Olives

CITRUS
Citrus sp.

Citrus fruits were one of the first fruits brought to the South. In Florida early settlers were planting citrus species by the end of the 1500s. Seedling trees were grown south of New Orleans in Plaquemines Parish around 1790. Texas' lower Rio Grande Valley citrus industry was started in the early 1900s. Since these early beginnings, citrus has become the leading southern commercial fruit crop. Florida's citrus industry produces over 70 million gallons of orange juice annually. Citrus is also a popular home crop for the lower South and areas near the Gulf of Mexico.

Climate

Freezes below 20°F can kill citrus trees; this limits citrus culture to the areas receiving 400 hours of chilling or less. The commercial citrus industries are limited to the 300-hour or less areas (Figure 10, page 13). But small orchards and specimen plantings of Satsumas can be successful along the Gulf Coast and should be tried more.

Citrus have no true rest period or chilling requirement and can grow any month of the year. Cultural practices have to be used to keep the trees dormant in December, January, and February.

Soil

Citrus can be grown in most soils: The Florida crops are primarily planted in sand, while Texas crops are grown in heavier soils. Dooryard citrus can be grown with as little as 1 foot of topsoil. Gulf Coast topography is usually flat and poorly drained, so trees should be planted on slightly raised rows to insure maximum surface drainage during the rainy season (Figure 42).

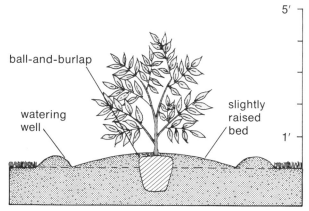

Figure 42. A balled-and-burlapped satsuma tree planted on a slightly raised bed with a watering well.

Crops and Varieties

Citrus crops and varieties are here divided into those which are most cold-hardy for the coastal region and those for the warmer commercial citrus regions.

Coastal Varieties. These are all grafted onto the trifoliate rootstock for maximum cold hardiness.

Satsuma (*Citrus unshiu*) is a very cold-hardy easy-to-peel mandarin which ripens early. It is an exclusively

Satsuma, a good, cold-hardy mandarin. This citrus fruit is commonly grown along the Gulf Coast.

coastal variety because it produces poor-quality fruit in the warmer commercial area. Its grafting onto the trifoliate rootstock produces a very productive, cold-tolerant dwarf tree which yields high-quality fruit. It will bear fruit in its second or third year and produce commercial-quality crops in its fourth.

Meyer lemon (*Citrus meyeri*) is a small, productive tree which is more cold-hardy than the common lemon. The fruit is medium-to-large and is produced throughout the year. It can remain on the tree for several months. Meyer lemon should be grafted onto trifoliate rootstock. The juice makes delicious lemonade and lemon pies.

Nagami kumquat (*Fortunella margarite*) produces small oval kumquats which have a delicious peel and very bitter flesh. The tree is small and very productive. The fruit color is a beautiful bright orange, contrasting with the dark green foliage. It should be grafted onto the trifoliate rootstock for maximum cold hardiness. Nagami kumquats make excellent yard trees and hedge rows.

Container-grown Satsuma, Meyer lemon, and kumquat trees are excellent patio specimens for the 400- to 700-hour chilling zone. Two or three times each winter, when the temperature drops below 28°F, the plants will need to be covered and moved indoors for cold protection. In the fall, the bright green foliage and orange fruit are simply beautiful. The blossoms are also very attractive and smell great for a couple of weeks each spring.

Warm-Climate Varieties. These are grafted onto the rough lemon rootstock in Florida, sour orange rootstock in the lower Rio Grande Valley, and the trifoliate in Plaquemines Parish, Louisiana.

Navel orange (*Citrus sinensis*) is a delicious large seedless orange which ripens in early December. The trees are slow to bear and have fruit-drop problems until they're fully mature. The navel tree will require good care for optimum production of high-quality fruit.

Valencia (*Citrus sinensis*) is a medium-sized round orange with few seeds. It ripens in February. The tree is large and very productive, and the fruit can store on the tree for a considerable period.

Ruby Red grapefruit (*Citrus paradise*) is the most common grapefruit grown in Texas. The fruit is large with deep pink flesh and delicious typical grapefruit taste. The tree bears at an early age and is very productive. Though there are several new, red-fleshed grapefruit varieties on the market, the Ruby Red is the best for home and small plantings because of its strong tree, productivity, and variable soil tolerance.

Dancy tangerine (*Citrus reticulata*) is a medium-sized, easy-to-peel, deep-orange-colored, late-ripening fruit which has many seeds. The tree is vigorous and productive.

Ponkan mandarin (*Citrus reticulata*) is a medium-sized, easy-to-peel, light-orange-colored, mid-ripening fruit which has few seeds. The flesh has a deliciously mild sweet taste. The tree is a vigorous, very upright grower which is not cold-hardy.

Temple (*Citrus temple*) is a medium-to-large-sized, relatively easy-to-peel, orange-colored, highly flavored fruit which has many seeds. The tree is moderately vigorous and is not cold-hardy.

Orlando tangelo (*Citrus sp.*) is a medium-sized, somewhat easy-to-peel, orange-colored fruit which has mild sweet orange-colored flesh and seeds when pollinated. The leaves of Orlando tangelo are distinctively cup-shaped. The tree is vigorous and productive when cross-pollinated.

Nagami kumquats have a delicious peel but bitter flesh. The trees are ideal as landscape specimens, hedge rows, and container plants.

Spacing

Your citrus site should have good air and water drainage. The deeper the soil, the better the tree will grow. If poor drainage is uncontrollable, plant the trees on ridges to aid in surface water runoff (Figure 42).

Citrus are usually planted on rows spaced 25 feet apart. Kumquats can be spaced 12½ feet in the row; kumquat hedges can be planted 6 feet apart. Satsumas, Meyer lemons, Dancy tangerines, Temple, and Orlando tangelos can be spaced 12½ feet apart in the row. Navel and Valencia oranges and Ruby Red grapefruits should be spaced 25 feet apart.

In the coastal area, don't plant citrus until the danger of freeze is past. Hold the trees in a protected area until March. This will reduce possible loss of young trees without previous conditioning or hardening-off.

Training and Pruning

Most citrus trees are purchased as ball-and-burlap or container trees and will not need to be cut back ½ at planting. If bare-root trees are planted, cut the top back ½ and remove all the leaves at planting.

Citrus require very little or no pruning for shaping. Rootstock suckers should be removed as they appear. They can be recognized by their upright growth, origin below the bud line, and thorns (Figure 43).

Fertilization

Apply 1 lb of complete 3-1-2 fertilizer per inch of trunk diameter. To reduce growth in September and October, don't fertilize after June. The tree should be fully dormant in December, January, and February to withstand potential freezes. This is especially important in coastal areas. If mature trees in the warmer areas are weak or heavily cropped, apply 4 lb of a complete fertilizer in August or early September.

Cold Protection

Wrap the trunks of 1- to 5-year-old trees with foam or fiberglass to insulate the trunk and bud union against freeze injury. The graft union and trunk can also be protected by mounding soil 2 or 3 feet over the crown. Some growers are going back to mounding rather than using foam insulation

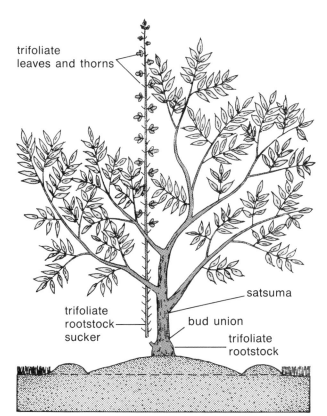

Before the first freeze, wrap the trunks of young citrus trees with foam rubber or soil to insulate them from the cold.

If you're in the coastal area, plant your trees on the south or east side of the house for additional wind and cold protection. The trees should be covered with boxes, garbage bags, tarpaulin or blankets if temperatures are expected to go below 25°F. This is extremely important in November or early December before the trees are fully dormant.

Figure 43. Remove rootstock suckers from satsuma trees.

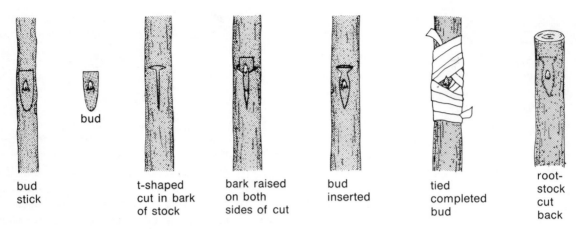

| bud stick | bud | t-shaped cut in bark of stock | bark raised on both sides of cut | bud inserted | tied completed bud | root-stock cut back |

Figure 44. Steps in "T"-budding citrus, peaches, and plums. After grafting, rootstock is cut back to force the bud to grow.

Good tree health is also important for maximum cold hardiness. Trees which are overcropped, mite-infested or drought-stressed are very susceptible to cold injury.

Propagation

Citrus are propagated by "T"-budding improved varieties onto seedling rootstocks (Figure 44). The seedlings are grown 1 year in special raised beds 2 inches apart. They are then lined out in the nursery row 6 to 12 inches apart. After one growing season in the nursery the rootstock seedlings are "T"-budded in September or February. The bud is then grown for 1 year before transplanting. Most citrus trees are sold with a 2-year root and a 1-year top. Occasionally, larger trees are sold which have a 2-year top and 3-year-old root.

Citrus Pests

Commercial citrus growers have to use a spray program to keep their fruit insect- and disease-free.

Mites are very tiny spiders which feed on the leaves and fruit. The rust mite is a very important citrus pest which can seriously weaken the trees and ruin the appearance of the fruit. The fruit will have a large number of very tiny lesions, giving it a "rusty" appearance. Mites are controlled with miticide sprays.

Melanose is a fungus that affects the quality of the fruit and damages the leaves. The fruit will have small raised dots on the skin surface. The entire fruit will have a brown appearance and be rough to the touch. The disease is prevented by fungicide sprays immediately after fruit-set.

Summer scale insects are occasionally a serious citrus problem. They are controlled by summer oil sprays marketed especially for citrus. Do not use dormant oil.

White flies are small insects which feed on the underside of the leaves. They are one of the most serious pests on homegrown citrus. Honey dew (sugary film) on the leaves is a sign of their presence. Sooty mold also develops on the honey dew, giving a black dusty appearance to the leaves. Summer oil sprays for scale will also control the white fly.

Harvest

Citrus are non-climacteric and should remain on the tree until fully ripe. The further north and colder the climate within the citrus-growing zone, the higher the quality of the fruit. The easy-to-peel crops such as Satsuma should be clipped, otherwise they will plug the peel.

COMMON FIGS
Ficus carica

The fig is one of the oldest fruit crops known to man, and it has long been an important home fruit crop in the South. In the early 1900s there was a 17,000-acre fig-processing industry on the Texas Gulf Coast, which attests to the fruit's adaptability in the South.

Figs can be grown as trees or bushes, depending on the way they are propagated and pruned (see pages 77–79).

Climate and Soil

Though the fig grows best south of the 800-hour chilling zone, it can be grown anywhere in the South. The tree is frost-sensitive and can receive occasional injury in all southern areas. If growth does not slow significantly in October, early freezes can kill the bush (or tree). However, mature bushes which are fully dormant can endure 10°F with little damage. Bushes should be planted on the south side of buildings in the colder areas of the South.

Most southern soils will grow healthy fig trees. Figs grow in sands or clay, high or low pH, and moderately drained soils. They are relatively salt-tolerant and can be grown in the Southwest or along the coast near brackish water.

Varieties

There are four types of figs: Caprifigs, Smyrna figs, San Pedro figs, and Common figs. Of these, only the Common fig is of significance to southern fruit growers. It is a seedless fruit which does not require pollination. The fruit is produced as a main crop on wood that has grown the same season. The fruit of southern fig varieties must have a closed "eye" to prevent entry by the dried fruit beetle.

Celeste (Malta) is a small, brown-to-purple fig which has a tight closed eye and a very sweet taste. The bush is vigorous, large, productive, and the most cold-hardy of the common fig varieties. The Celeste bush should not be pruned heavily, for this can reduce the crop.

Texas Everbearing (Southern Brown Turkey) is a medium-sized, light brown fig which has two crops. The first is of large, light-colored fruit, called Breba figs, borne on last year's wood. The second is medium-sized, brown fruit on the current season's wood. By thinning out 25 to 33 percent of the shoots each winter, the tree will always produce vigorous 1-year-old shoots that bear the Breba crop. This heavy pruning will also increase freeze injury, which limits Breba crop production to within 100 miles of the Gulf of Mexico. The eye of the fig is moderately closed, which helps reduce fruit spoilage on the tree. The bush is very vigorous, large, and productive and the fruit has a mild, sweet flavor. Everbearing figs grow and yield best in the cooler areas of the South.

Figs are sweet fruit; however, when young they are frost-sensitive and should be grown only south of the 1,200-hour chilling zone.

Alma is a new Texas A&M University fig which is medium-sized with a cream-colored peel. The fruit is extremely sweet and delicious. The bush is moderately vigorous, medium-sized, and extremely productive. The eye of the fruit is sealed with a drop of thick resin which inhibits on-the-bush fruit spoilage. Alma is frost-sensitive as a young bush, which limits its culture to the area south of the 800-hour chilling zone.

Other Varieties. The *Magnolia, Brunswick, Mission*, and *Kadota* varieties are not recommended because they have an open eye. The *Hunt* variety is a good southern fig but is extremely freeze-sensitive and is not readily available.

Spacing

Space fig bushes 12 to 20 feet apart, and plant them several inches deeper than they grew in the nursery row. This protects stem tissue below the soil surface from winter injury. If the bush freezes to the ground line, stem tissue below the surface can develop shoots.

Propagation

Figs can be propagated by suckers, layering, or cuttings. Using suckers from the crown of a mature bush is not recommended because it will transfer nematodes from the roots of the mother bush. Air layering is occasionally used to propagate new plants, but it requires special techniques and practice.

The easiest way to propagate figs is by stem cuttings. First, collect 6- to 8-inch terminal shoots of healthy 1-

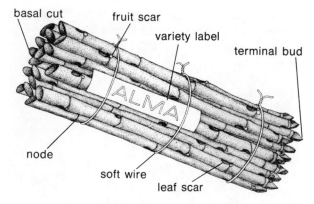

Figure 45. A bundle of fig stem cuttings.

year growth in late winter. (Thin stems make weak cuttings and should not be used.)

Next, group the cuttings in bundles for callusing (Figure 45). Invert the bundles in a callusing trench from mid-January to mid-April, covering the basal ends of the cuttings with approximately 2 to 4 inches of soil (Figure 46). The callusing trench should be well drained and weed-free.

After callusing, place the cuttings right side up in a propagation row (Figure 47). Place the cuttings with 1 inch above the ground and 6 inches of stem below the ground, and space them 6 to 12 inches apart in the propagation row.

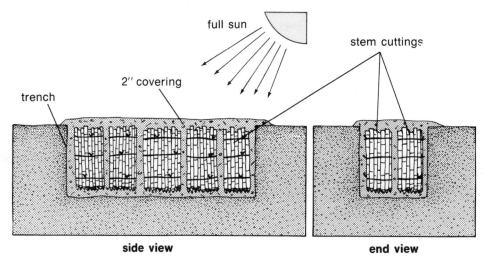

Figure 46. Callusing trench with upside-down bundles of fig or grape stem cuttings.

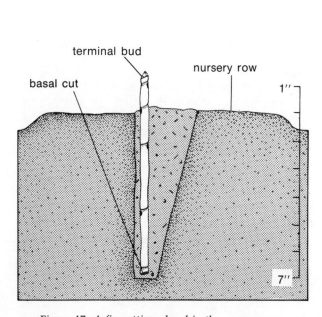

Figure 47. A fig cutting placed in the nursery row.

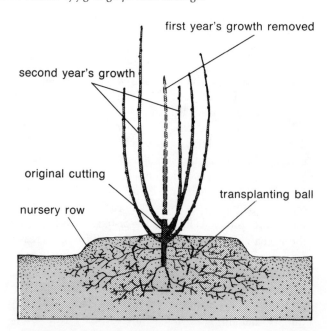

Figure 48. A two-year nursery fig bush ready to be balled-and-burlapped.

The cuttings will develop roots and make good growth in just 1 year in the propagation row—shoots can make 36 to 48 inches of growth in 1 year. The small tree will be ready for transplanting during the dormant season.

Large bushes can be produced by leaving the plant in the propagation row for a second season. Multi-shoots are stimulated by cutting off the first year's growth at the ground level during the dormant season. Figure 48 illustrates a multi-shoot 2-year-old nursery bush in the propagation row. These bushes must be ball-and-burlap transplanted.

Training and Pruning

Figs can be trained to a single-trunk open vase-type tree or to a multi-trunk bush. The bush system is by far the best for the South because freezes occasionally kill the upper part of the plant. The tree system can be used only in the warmer 200-hour chilling zone. Figure 49 illustrates the two types of training.

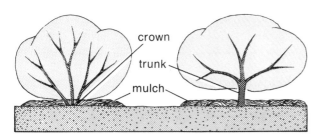

Figure 49. Mature figs trained to a bush and tree.

Normally, figs are pruned very little. Mature Celeste and Alma trees should not be pruned since this will reduce the crop size. Everbearing will produce a fair crop following heavy winter pruning. Dehorning figs encourages rapid growth and will increase water stress.

Older trees which make little growth each year should be thinned out to stimulate new growth. This will also increase fruit size. The trees should be pruned enough to stimulate approximately 1 foot of growth each year. All weak, diseased, or frozen limbs should also be removed each dormant season. Frozen limbs should be thinned out after damage becomes obvious in late spring.

Fruit Drop and Water Management

Special attention has to be given to soil moisture management in growing figs to prevent fruit drop. Most of the fig tree roots are close to the soil surface and can eas-

ily dry out. You should assume your figs have root-knot nematodes and manage them accordingly.

For these reasons, water must be applied to the trees as drought develops. Slight leaf wilting in the afternoon is a good indication of water stress, which causes premature fruit drop in common fig varieties. Mulching with straw or grass clippings will help maintain soil moisture and reduce weed competition for available water.

This fruit drop problem is very common in hot, dry areas when the fig tree is grown in shallow soil and the roots are infested with nematodes. Common figs do not have true seeds to produce hormones which prevent fruit drop naturally.

To water properly, apply 7 gallons of water per bush per week for each year of age, up to 35 gallons per week. Reduce watering in August to help the tree go dormant. A mature bush which has lost all its leaves and has become totally dormant can withstand much cooler temperatures than one which is rapidly growing at the time of first frost. Stop watering in September to reduce growth and encourage the onset of dormancy.

Fertilization and Mulching

Organic mulches such as grass clippings, hay, or pine needles are extremely important in growing healthy fig trees. Mulch the bush 12 inches deep. The mulch will insulate warm soil temperatures in the winter and prevent the crown of the bush from freezing; it will conserve soil moisture, cool the soil, and control weeds in the growing season. And the mulch will organically supply the nutrients—don't use commercial fertilizers on figs as it will stimulate excessive growth.

Mulch your fig bushes to insulate the soil during winter.

Nematodes on the roots of a fig bush. Notice the knotted areas on the roots.

Fig Pests

Nematodes are small microscopic worms that are a universal fig problem. Figs seldom are without nematode infestations. They feed on small roots, reducing movement of nutrients and water within the roots. For this reason, figs should receive optimum moisture management. This includes regular watering, no heavy pruning, no commercial fertilization, and excellent mulching for water conservation and weed control. To prevent or delay the onset of nematodes, always use stem cuttings—not suckers—for propagating new plants. Never plant figs in an old garden site which contained tomatoes. Always inspect the roots of new plants to ensure they are not infected with nematodes (small knots, see photo).

Harvest

Harvest figs as soon as they are ripe; they are non-climacteric and will not ripen after harvest. If overripe fruit remains on the bush for more than a day, birds, dried fruit beetles, and spoilage will claim the crop. Never drop spoiled fruit under the bush because they will attract more insects. On-the-bush spoilage is usually the result of microorganisms carried into the fig eye by the dried fruit beetle. This is why closed-eye fig varieties are recommended.

Wear gloves, a long-sleeved shirt, and a scarf when harvesting figs to prevent skin irritations from the white fig latex or sap. Do not eat green immature figs, because excess ficin in the white latex of the fruit can damage the mucus lining of the mouth.

Fully ripe fruit are soft, the stem and skin separate easily from the flesh, and no white latex is present. Fully ripe fruit usually have cracks in the skin from the rapid swell in fruit size during ripening. Many people will not eat a fig because they had previously eaten figs before they were ripe; however, when fully ripe, figs have a very delicious, high-sugar, sweet taste. They also make excellent preserves.

ORIENTAL PERSIMMONS
Diospyrus kaki

The oriental persimmon is a small, easy-to-grow, warm-climate tree, which is well adapted to all areas of the South. It cannot be grown north of the 1,000-hour chilling zone. The common persimmon, *Diospyrus virginiana*, grows wild throughout the South and is discussed in greater detail on page 84.

The persimmon will grow in a very wide range of soils—once established, the trees perform well in both sandy and clay soils. One of the persimmon's outstanding characteristics is its freedom from major insect and disease problems.

Mature trees can reach a height of 40 feet, while some varieties remain less than 10 feet. They produce prolific

The fruit of the Hachiya persimmon. This outstanding tree is relatively insect- and disease-free, and it makes a lovely ornamental in the landscape while providing you with delicious fruit. The bright orange fruit is produced in the fall, giving the yard some additional color.

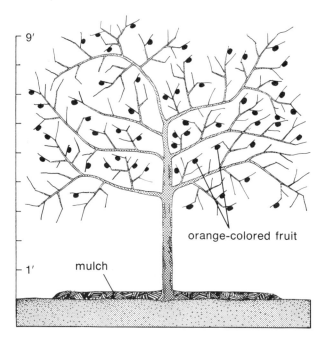

Figure 50. A typical mature dormant Eureka oriental persimmon tree.

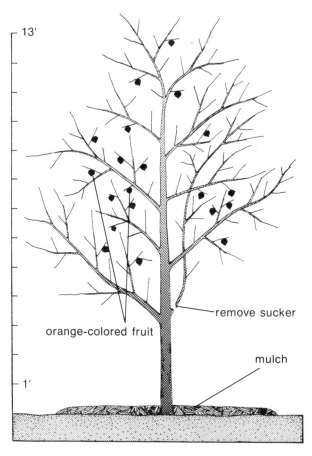

Figure 51. A typical mature dormant Hachiya oriental persimmon tree.

crops of very attractive fruit during the fall when few other fruits are ripening. The fruit is very delicious when fully ripe and is high in vitamin A.

Varieties

The first improved persimmon varieties were introduced into the U.S. in the late 1800s from the Orient, where they had been cultured for several hundred years. In China the fruit is eaten both fresh and dried.

Eureka is a heavy-producing, medium-sized, flat-shaped, extremely high-quality, red persimmon. The tree is relatively small and is self-fruitful (Figure 50). The fruit contains seeds; however, seedless fruit can be obtained (see "Pollination"). Eureka has proven to be the best commercial variety in Texas.

Hachiya is a productive, large, cone-shaped, seedless persimmon with bright orange-red skin. The tree is vigorous and upright (Figure 51). Hachiya is an outstanding variety; it makes an excellent dual-purpose fruit-ornamental specimen.

Tane-nashi is a moderately productive, cone-shaped, orange-colored persimmon. The tree is vigorous and upright. The fruit stores extremely well on the tree and is seedless. Tane-nashi makes an excellent landscape ornamental.

Tamopan is a moderately productive, very large, orange, flat-shaped persimmon with a distinctive ring con-

striction near the middle of the fruit. The tree is the most vigorous and upright of the varieties grown in the South.

Fuyu (Fuyugaki) is a medium-sized, nonastringent, self-fruitful persimmon. The fruit is red, rather square-shaped, and can be eaten green. Fuyu is self-fruitful and will pollinate all other varieties.

Pollination

Oriental persimmons frequently fail to produce full crops. This is due in part to pollination and environmental stress problems.

Male, female, and/or perfect flowers can be produced on the same tree on current season's growth. Hachiya, Tane-nashi, and Tamopan produce flowers which develop into excellent parthenocarpic (non-fertilized) fruit without pollination or seeds. These varieties can be pollinated by common persimmon or Fuyu to produce fruit with seeds. Some varieties will have darker fruit flesh when seeds are present; this doesn't occur with Hachiya, Tane-nashi, and Tamopan.

Fruit drop is common on fruits producing parthenocarpic fruit without true seeds. Any stress problems such as excessive heat, drought, cold, or flooding can stimulate fruit drop. Therefore, mulching and good water management are advised.

Propagation

Japanese persimmons are grafted onto common American persimmon rootstocks for strong trees and adaptability. The whip-and-tongue graft (see page 54) on 1/4-to 1/2-inch diameter trees at the ground line is the most successful and commonly used method.

Common American persimmon seeds should be planted soon after harvest in the fall or placed immediately into a moist medium for stratification at 45-50°F. In January or February the stratified seeds should be planted in nursery rows. The seedlings are grown 2 years in the nursery before whip grafting in January or February. The grafts of the improved persimmons are then grown in the nursery row for 1 year. The trees can be transplanted bare-root in January, February, or March. They usually bear in 3 to 4 years.

Harvesting

The persimmon is a climacteric fruit and will ripen after being picked from the tree. It is very astringent until fully ripe.

Ripening in the South is usually associated with the first fall frost. The fruit can be frozen in a refrigerator overnight to accelerate the loss of astringency. (The Fuyu variety can be eaten green with no astringency.) Persimmons will store on the tree for a considerable period into the winter, and the tree and fruit are very attractive during this period.

OLIVES
Olea europola

Olives are excellent, non-fruiting ornamentals south of the 700-hour chilling zone; however, they freeze to the ground 10 percent of the time in the cooler areas of this zone. Jim Denney, while at Texas A&M, learned that the olive can fruit in the 400- to 600-hour chilling zone. He showed that the olive sets fruit buds in late winter, eight to ten weeks before bloom, in response to both cool nights and warm days. Trees grown in this zone could fruit, though not necessarily as a commercial crop. The olive would make an excellent addition to grounds and gardens associated with churches or religious buildings. Since freeze is always possible, olives should be planted in an area protected from north winds. The varieties recommended are *Ascolano* because of its cold hardiness, *Manzanillo* and *Missiore* because of their availability at nurseries, and *Barouni* because it originated in a warm climate and may require fewer cool periods for fruit-bud development.

Wild Fruits, Berries & Nuts and Other Crops

Wild Fruits, Berries & Nuts

Many cultivated fruit crops originated in the South in the wild, and they continue to grow along rivers, streams, roadsides, fence rows, and in the forest, totally uncultivated in any way. These fruits are truly organic, with no direct interference from man. They are eaten and enjoyed by both wildlife and man. We hope that they will always be with us. Though there are many more wild crops than those mentioned here, those listed here are the most rewarding, both in quality and quantity.

NATIVE PECANS
Carya illinoensis

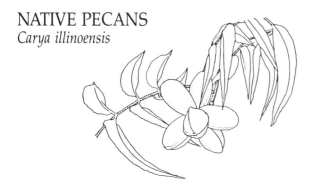

Native pecans cover more than 1 million acres of land along the banks of the creeks and rivers of Texas, Oklahoma, Louisiana, and Arkansas. Their small nuts have a better taste than the larger, improved varieties. The trees tend to alternate bear, producing very high tonnage one year, followed by little or no production for 1 or 2 years. These pecans can be very small, with more than 700 nuts per pound. Squirrels, deer, crow, raccoons and turkey feed heavily on the native pecan. Native pecans were a major food for the Indians of the Southwest dur-

ing the "on" years, or the years of heavy pecan production. Shuck opening usually begins in late October and harvesting can continue through January. Many of the improved varieties were derived from native pecans selected from the wild.

WILD PLUMS
Prunus angustifolia

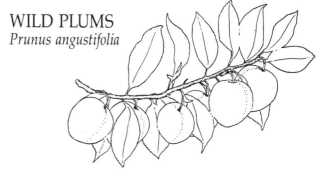

Wild plums grow in thickets all across the South. There can be 1,000 small trees growing on less than an acre of land. The trees grow in full sun and are very small (less than 8 feet tall) with bushy tops and reddish-brown bark. Make sure you have gloves when you find them, because the twigs can have thin thorns. The patches, or thickets, form a beautiful white blanket of blossoms in the early spring, and the fruit ripens in late May. Indians evidently transplanted them throughout the South as semi-cultivated plants. In some areas the wild plum is called the Chickasaw plum.

The fruit varies from yellow to dark red when fully ripe, and from egg-shaped to round, with a very thin skin and soft, very sweet taste. Wild plums tend to alternate bear, with a very heavy crop one year and a wait of several years before another. They bloom early; consequently, the blooms are subject to frost damage. Wild plums are delicious when eaten fresh, and also make excellent jelly.

COMMON PERSIMMONS
Diospyrus virginiana

Common persimmons thrive in abundance all across the South and are a favorite early-winter snack for hunters and hikers. These small, dark orange fruit are sweet and delicious when fully ripe. As they mature the skin turns black and all of the astringency disappears. This usually occurs after 2 or 3 freezes. Everyone remembers his first bite into a green persimmon; your mouth is sucked totally dry and it seems impossible to talk. Strangely, the same fruit is soft, juicy, and delicious when fully ripe.

The trees can grow to 40 feet in height, and sprout readily from roots forming persimmon thickets in pastures or along fence rows in full sun. Possum, raccoons, and deer feed heavily on wild persimmons. The tree is an excellent rootstock for the cultivated oriental persimmon, but it can be difficult to graft.

MAYHAWS
Crataegus aestivalis
and *C. opaca*

Mayhaws are very common south of the 1,000-hour chilling zone. They grow under hardwood timber in the wet floodplain soils along creeks and rivers. These small trees are of the Hawthorne family. They have a low-chilling requirement and beautiful white flowers in the early spring.

The fruit is similar to a very small, highly acidic apple and makes outstanding jelly. The fruit ripens in early May, falls into the nearby water, and moves downstream. It can be harvested by someone shaking the trees, so that the fruit falls into the water, and someone else trapping the floating fruit downstream. As you might expect, snakes are a constant danger when harvesting mayhaws. Nevertheless, in the old days, most southern rural families made a big batch of mayhaw jelly every year. The jelly is especially good when spread onto hot, buttered biscuits or white cake. Outstanding

seedlings have been selected and propagated in recent years. These varieties include *Lori, Lindsey, Big Red, Super Spur, Mason Super Berry, Highway Super Berry,* and others.

SOUTHERN DEWBERRIES
Rubus trivialis

Southern dewberries are as southern as you can get. These small, early-ripening berries grow on perennial canes along every railroad, highway, or abandoned field in the South. They make an excellent jelly that is distinctly different from common blackberry jelly. If you decide to pick dewberries, be prepared to fight the snakes and red bugs. A good stick, rubber boots, and lots of insect repellent are essential. Dewberries are low growers and require a lot of bending over, but the effort is worth it because dewberry jelly is among the best.

BLACKBERRIES
Rubus allegheniensis

Blackberries grow wild throughout the South, and were made famous in the story of Br'er Rabbit in the briar patch. They begin to ripen right after the dewberries and can be purchased along the roadside by the gallon. Unfortunately, they ripen all at once and are only available for a couple of weeks. The development of new, highly-productive, large blackberries has shifted many people away from the wild native fruit.

WILD MUSCADINES
Vitis rotundiflora

Muscadines grow wild throughout southern forests. The vines frequently climb over 100 feet to the top of the trees. Only one out of every 10 vines is a female and bears fruit. The fruit is red-to-black and ripens in the fall, one berry at a time, in a cluster. The skin is very tough and the flesh has a musky, wild flavor. The muscadine can be eaten fresh or made into delicious jelly. In years past, most southern rural boys had a favorite muscadine tree they could climb up into and relax, graze on the fruit, and be isolated from their work in the sanctuary of the tree.

MUSTANG GRAPES
Vitis candicans

Mustang grapes grow wild along almost every road and stream in central Texas. The vine is very vigorous and is drought and heat tolerant. It produces very acid fruit in small clusters that ripen in late July. These are harvested green for green grape pie, or fully ripe for Mustang wine. In the old days, and to a great extent today, many German and Czechoslovakian families in Texas had someone who was really good at making Mustang wine. The grapes were harvested when fully ripe, mixed with two parts sugar and water for each part grapes, and fermented in oak barrels until Christmas when the first samples were taken. Since European grapes did not survive in the new world, the settlers had to turn to the native Mustang for wine. One day a market may develop for this brilliant red, very sweet, acid, tannic wine.

Other Crops

In addition to the principal fruit crops discussed thus far, many lesser fruit crops are popular across the South. Homeowners have successfully grown the following fruits for many years.

LOQUATS
Eriobotrya japonica

The loquat is a very productive small yard tree along the Gulf Coast and in southern Florida and Texas. The tree is an evergreen, has few insect and disease pests, and makes an excellent ornamental. It is related to the apple, pear, and quince and is susceptible to fire blight.

The fruit is oval-shaped and borne in clusters. It can be eaten fresh or made into jelly. The main varieties include *Early Red, Oliver, Advance,* and *Champagne.* The loquat is a low-maintenance tree requiring little or no fertilizer, pruning, or irrigation.

POMEGRANATES
Pumica granatum

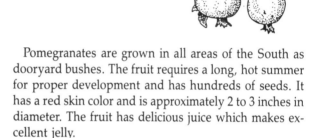

Pomegranates are grown in all areas of the South as dooryard bushes. The fruit requires a long, hot summer for proper development and has hundreds of seeds. It has a red skin color and is approximately 2 to 3 inches in diameter. The fruit has delicious juice which makes excellent jelly.

The bush, composed of many canes which grow to a height of 10 feet, has few pest problems. The pomegranate can be grown quite well in the Southwest, where alkaline soil and high salt limit other fruits. It is a very popular fruit in much of Mexico. The *Wonderful* is the main variety propagated.

Pomegranates can be grown as yard bushes. The plant is adaptable to alkaline soils and high salt concentrations, and its fruit makes a delicious jelly.

Jujubes, an edible fruit and an interesting tree for the landscape.

AVOCADOS
Persea americana

Avocado seeds can be planted directly from the store-bought fruit for ornamental pot plants. Don't let the seed dry out, and plant it large end down with the tip of the seed slightly exposed. These pot plants seldom bear fruit and must be moved indoors if freezes occur.

JUJUBES
Ziziphus jujuba

Avocados are large evergreen trees which are very cold-tender and are limited to southern Florida, Plaquemines Parish, Louisiana, and south Texas. Temperatures below 25°F will cause some damage. There are three races of avocados. *Lula,* a hybrid of the Guatemalan and West Indian races, is the most widely propagated variety. Mexican race hybrids have good cold tolerance; however, they are very susceptible to anthracnose. Trees killed to the ground can regrow rapidly and produce in 4 years. Avocados are very climacteric fruit and will ripen rapidly after harvest.

Jujubes are small, heat- and drought-tolerant trees which grow to a height of 30 to 50 feet in all areas of the South. They produce small, egg-shaped fruit which ripens in August and has a peel and flesh similar to an apple's and a seed similar to an olive's. The *Lang* and *Li* are the only propagated varieties grown. The tree is very unusual, with rough bark, shiny leaves, and angular limbs. It is a very attractive landscape tree.

Fruit Management References

Southern Fruit Nurseries

It is always best to buy fruit trees from a well-known local nurseryman. If he can't supply the varieties recommended, contact the following mail-order nurseries.

(Space limits a complete list of Southern nurserymen. No discrimination is intended and no guarantee of reliability is implied.)

Adams Citrus Nurseries Inc.
2020 Dundee Road
Winter Haven, Florida 33880

Armstrong Nursery
P.O. Box 473
Ontario, California 91761

Auburn Nursery
Auburn, Alabama 36830

Bountiful Ridge Nursery
Princess Ann, Maryland 21853

Chestnut Hill Nursery
Route 1, Box 341
Alachua, Florida 32615

Conner Strawberry Nursery
Augusta, Arkansas 72006

Cumberland Valley Nurseries, Inc.
McMinnville, Tennessee 37110

Fincastle Nursery
Larue, Texas 75770

Finch Blueberry Farm
Bailey, North Carolina 27807

Haley's Nursery
Smithville, Tennessee 37166

Hilltop Nurseries
Hartford, Michigan 49057

Hollydale Nursery
Pelham, Tennessee 37366

Ison's Muscadine Nursery
Brooks, Georgia 30205

J.E. Leger
Blueberry Nursery
Ocilla, Georgia 31775

Lewis Strawberry Nursery
Rocky Point, North Carolina 28457

Linwood Pecans
Turlock, California 95380

Mandarin Berry Farm
11243 Scott Mill Road
Jacksonville, Florida 32217

Mountain Creek Nursery
Route 5, Box 112-A
McMinnville, Tennessee 37110

Owen's Muscadine Nursery
Gay, Georgia 30218

Pape Pecan House
Box 1281
Seguin, Texas 78155

Powell Blueberry Nursery
Thomasville, Georgia 31792

Southeast Nurseries, Inc.
Route 1, Box 321-A
Raleigh, North Carolina 27609

Southern Orchards, Inc.
Rt. 5, Box 73B
Louisville, Mississippi 39339

Stark Bros.
Nurseries & Orchard Company
Louisiana, Missouri 63353

Stribling's Nurseries, Inc.
P.O. Box 793
Merced, California 95340

Sunsweet Berry & Fruit Nursery
Box D
Sumner, Georgia 31789

Texas Pecan Nursery
P.O. Box 306
Chandler, Texas 75758

Wells Nursery
P.O. Box 146
Lindale, Texas 75771

Womack's Nursery
DeLeon, Texas 76444

Spray Schedule for Southern Fruits

This guide provides information on insect and disease problems of peaches, plums, and pecans.

Homeowners should be familiar with pests and diseases, their life cycles, and damage. Problems must be identified and the proper control methods selected. The situation is often complex because such problems vary from one area to another and with the time of year. Keep records of pest and disease occurrence to assist in making control decisions.

Cultural Practices

Healthy plants are less susceptible to insect and disease attack. Optimum tree growth is maintained by following a well-balanced fertility program, selecting adapted disease-resistant varieties, and practicing proper pruning and other cultural practices.

Proper clean-up and plant residue disposal are important in reducing plum curculio, hickory shuckworm, brown rot of peach, and pecan scab.

Pesticide Safety

Before using any pesticide, carefully read all instructions on the container. Note any special directions such as the need to wear protective clothing. Take necessary precautions when applying pesticides to avoid unnecessary chemical exposure.

Mix pesticides in a well-ventilated area or outdoors. Avoid chemical contact with the skin and do not breathe chemical vapors.

Apply pesticides at the proper rate. Using less chemical than prescribed may result in poor control, while using more than recommended may result in excessive residue on the fruit or plant damage. It is also a waste of money.

Store chemicals in a secure area away from pets and children. Prepare only the amount required for one application. Properly dispose of any unused, diluted sprays and empty pesticide containers. Never store pesticides in unmarked jars, cans, or bottles.

Taken from Fact Sheet L-1140, *Homeowner's Fruit and Nut Spray Schedule,* by Jerral D. Johnson and Charles L. Cole, Texas Agricultural Extension Service.

Pesticides should be applied only by a certified applicator and according to label instructions. Chemical names used here are for educational purposes only and no endorsement or discrimination is intended.

Homeowner's Spray Schedule for Pecans

Timing	Pest	Pesticide	Rate/1 gal. water	Remarks
Dormant season (winter)	*Insects* scale insects phylloxera	97% oil emulsion	1/4 pt.	Spray tree trunks and branches thoroughly.
Budbreak (just as the buds begin to split and show green color). Terminal bud growth should be 1 to 2 inches in length.	*Nutritional* rosette	Zinc sulfate 36% WP or Zinc nitrate 17% liquid or Tracite-N-Zinc® 17% liquid or NZN® 6% liquid or ZN Special® 13.5% WP	2 tsp. 2 tsp. 2 tsp. 2 tsp. 2 tsp. 2 tsp.	Zinc sprays are essential for early season pecan growth. Early frequent applications will give the best response. Elemental zinc is toxic to most plants other than pecans and grapes; therefore, avoid drift to protect from phytotoxicity. If drift is a possibility, use NZN. Do not use any zinc product at higher than labeled rates since foliage burn can result. When applying more than one zinc spray in 2 weeks, reduce rate by one-half. Never spray young trees that are not actively growing.

(schedule continued on next page)

Homeowner's Spray Schedule for Pecans *(continued)*

Timing	Pest	Pesticide	Rate/1 gal. water	Remarks
	Insects phylloxera	Malathion® 50% EC	2 tsp.	If dormant oil was not used, then treat trees where a history of phylloxera damage indicates a need for control.
	Diseases scab and other foliage diseases	Benomyl 50% WP or Thiophanate-methyl 70% WP	1½ tsp. 1½ tsp.	
Prepollination (when leaves are one-third grown and before pollen is shed)—mid-April	*Nutritional* rosette	Same as bud-break		
	Diseases scab and other foliage diseases	Same as bud-break		
	Insects fall webworm, walnut caterpillar	Dipel® or Diazinon® 25% EC or Malathion® 50% EC or Zolone® 34.4% EC or Guthion® 12% EC or Sevin® 50% WP	2 tbsp. 2 tsp. 2½ tsp. 1¼ tsp. 1⅓ tbsp. 2 tbsp.	Repeat sprays as pest problem reoccurs.
Pollination (when casebearer eggs appear on tips of nutlets)—May	*Nutritional* rosette	Same as bud-break		
	Insects pecan nut casebearer	Same as pre-pollination		Apply in late April or during May (consult your county Extension agent for precise local timing).
	Diseases scab and other foliage diseases	Same as bud-break		

(schedule continued on next page)

Homeowner's Spray Schedule for Pecans *(continued)*

Timing	Pest	Pesticide	Rate/1 gal. water	Remarks
Second generation casebearer (42 days after first casebearer spray)	*Insects* pecan nut casebearer	Same as pre-pollination		
	aphids	Diazinon® 25% EC or	2 tsp.	Treat yellow aphids where an average of 20 per compound leaf is found or when excessive honeydew is produced.
		Malathion® 50% EC or	2½ tsp.	
		Zolone® 34.4% EC or	1¼ tsp.	
		Guthion® 12% EC or	1⅓ tsp.	
		Dimethoate 257® 30.5% EC	0.5 tbsp.	
	Diseases scab and other foliage diseases	Same as bud-break		May be required during extended periods of high humidity.
Water stage (when water collects inside nut)—mid- to late July	*Diseases* scab and other foliage diseases	Same as bud-break		Treat when there is a history of disease problems and during extended periods of high humidity.
Half-shell hardening—mid- to late August	*Diseases* scab and other foliage diseases	Same as bud-break		
	Insects aphids	Same as second generation casebearer		Treat yellow aphids when an average of 10 per compound leaf is found or when excessive honeydew is produced and aphid populations persist.
	hickory shuckworm	Zolone® 34.4% EC + Guthion® 12% EC	1¼ tsp. 1⅓ tbsp.	
	pecan weevil	Sevin® 50% WP	2 to 3 tbsp.	Treat areas with a history of pecan weevil infestation. One to three treatments at 10- to 14-day intervals are needed for heavy weevil infestations.
	mites	Guthion® 12% EC	1⅓ tbsp.	Apply when mites appear on foliage.

WP = wettable powder
EC = emulsifiable concentrate

Homeowner's Spray Schedule for Peaches and Plums

Timing	Pest	Pesticide	Rate/1 gal. water	Remarks
Dormant season	*Insects* scale insects	97% dormant oil	¼ pt.	Apply when temperature is between 40 and 70°F. Use only once.
Bud swell	*Disease* peach leaf curl	Chlorothalonil 40.4% F	1 tsp.	Apply if fall copper or chlorothalonil applications were not made.
Pink bud (just before bloom when buds show color)	*Disease* brown rot	Captan® 50% WP or Sulfur 97% WP or Benomyl 50% WP or Thiophanate-methyl 70% WP or Chlorothalonil 40.4% F	4 tsp. 7 tsp. 1 tsp. 1 tsp. 1 tsp.	
Petal fall (when flower petals begin to fall)	*Insects* plum curculio	Malathion® 50% EC or Sevin® 50% WP or Zolone® 34.4% EC	5–6 tbsp. 2 tbsp. 1¼–2½ tsp.	Apply when 75% of petals have fallen, and there is a history of insect damage.
	peach twig borer	Diazinon® 25% EC	2 tsp.	The peach twig borer usually is a problem only in West Central Texas.
	lesser peach tree borer	Thiodan® 9.7% EC	2 tbsp.	Make two applications approximately 3 weeks apart. Thoroughly wet tree limbs with spray.
	Diseases scab	Chlorothalonil 40.4% F	1 tsp.	
Shuck split (when the calyx separates from base of newly formed fruit)	*Insects* catfacing insects, plum curculio	Same insecticides as petal fall		Treat when there is a history of catfacing insects and/or plum curculio.
	Diseases scab	Same fungicide as petal fall		

(schedule continued on next page)

Homeowner's Spray Schedule for Peaches and Plums *(continued)*

Timing	Pest	Pesticide	Rate/1 gal. water	Remarks
First cover (treating newly exposed fruit)	*Insects* catfacing insects, plum curculio	Same insecticides as petal fall		
	Diseases scab	Same fungicide as petal fall		
Second cover	*Insects* catfacing insects	Same insecticides as petal fall		
	Diseases scab	Captan 50% WP or Sulfur 97% WP	4 tsp. 7 tsp.	
Third cover	*Diseases* scab	Same fungicides as second cover		
Pre-harvest	*Diseases* brown rot, rhizopus rot	Benomyl 50% WP Thiophanate-methyl 70% WP	1 tsp. 1 tsp.	
Post harvest—mid- to late August	*Insects* peach tree borer	Chlorpyrifos 12.9% Lindane® 20% EC Thiodan® 9.7% EC	2 tbsp. 1 tbsp. 2 tbsp.	Thoroughly wet from base of tree up to first scaffold limbs.
October 15 to December 1	*Diseases* peach leaf curl	Kocide 101® 77% WP or Chlorothalonil 40.4% F	2 tsp. 1 tsp.	Apply when leaves are falling from tree. Spray to run-off. Sprays are applied between Oct. 15 and Dec. 1.

WP = wettable powder
EC = emulsifiable concentrate
F = flowable

Family Orchard Planner

(For details, see pages 2-6)

Location _____ Date _____

Crop _____ Varieties _____
Crop _____ Varieties _____
Crop _____ Varieties _____
Crop _____ Varieties _____
Crop _____ Varieties _____
Crop _____ Varieties _____
Crop _____ Varieties _____
Crop _____ Varieties _____
Crop _____ Varieties _____

Crop Total _____ Variety Total _____

Soil:

Texture _____ Soil pH _____
Depth of topsoil_____ Type subsoil _____
Surface drainage _____ Internal drainage _____

Climate:

Record minimum temperature _____
Winter chilling received _____
Average last spring frost _____
Average first fall frost _____
Number of growing days _____

Site:

Frost pocket _____ Highway _____
Animal problems _____ Tax problems _____
Disease history _____ Available water _____

Orchard Plans:

Design _____
Spacing _____
Plant source _____
Tree delivery date _____
Ground preparation _____
Irrigation _____

Marketing Plans:

Family use _____
Roadside retail _____
Wholesale _____

Guidelines for Pruning Fruit Trees

Each fruit crop requires specific pruning and training; however, there are basic principles which are relevant to all pruning and they are covered here (Figure 52).

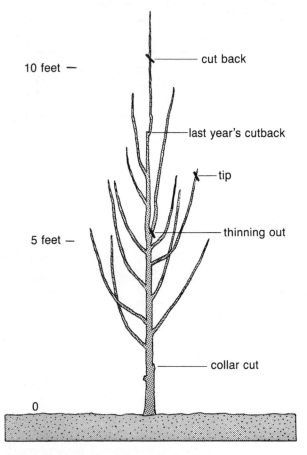

Figure 52. Examples of the various pruning cuts on a three-year-old tree.

TYPES OF PRUNING

Cut Back or *Heading Back* is when a shoot, limb, or trunk is cut back significantly. As much as 50 percent of the shoot can be removed.

Thinning Out is when an entire shoot or limb is removed, leaving only a collar cut at the trunk or limb.

Tip Pruning is when only the apical bud or bud cluster is removed from a dormant limb. It usually requires a pole pruner and only the very tip is removed. If more than 5 percent of the shoot is removed, it is considered heading back.

Pinch Pruning is the removal of a shoot apex or tip during the growing season. It is usually accomplished by using your fingers rather than pruning tools.

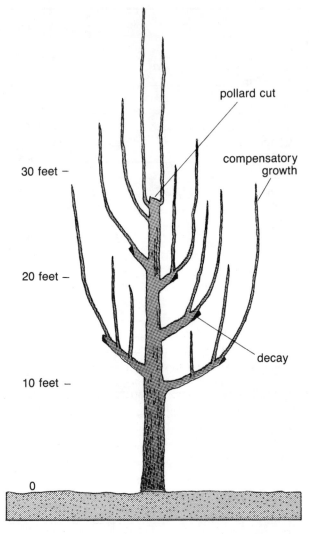

Figure 53. Mature dehorned tree. This type of pruning is not recommended.

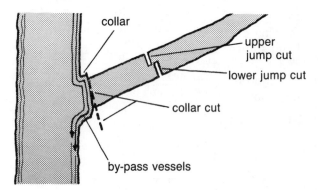

Figure 54. Proper limb pruning technique using a collar cut.

Hedging is the removal of a side or top of a canopy using power hedging equipment, in the dormant or growing season. The plant is usually maintained at a desired size and shape by frequent hedging.

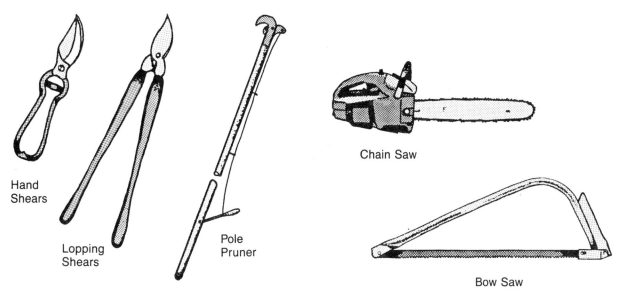

Figure 55. Pruning tools.

Dehorning or *Pollarding* is the removal of large limbs, leaving short, thick stubs (Figure 53). This type of pruning is very detrimental to the tree and is not recommended. Vigorous shoots, called compensatory growth, result following dehorning. This growth is non-fruitful and weaker than the shoots that were dehorned. Decay frequently enters the sawed-off stubs.

Tree Removal is the cutting down of a tree at the ground line. It is one of the best but least frequently used methods of correcting tree crowding.

Collar Cut. This is the best sytem for thinning out a limb (Figure 54). As food moves down a trunk, special by-pass vessels circle limbs which join the trunk. These by-pass vessels form a collar, or slight enlargement of the limb diameter, adjacent to the trunk. Do not remove the collar when thinning out a limb. If the collar is cut off, food will not be able to reach the tissue below the cut and decay will set in. If the collar cut is made properly the wound will heal or grow over much faster. Therefore, do not cut the limb off even with the trunk.

Jump Cut. Always cut a limb off by using three cuts; first, a lower jump cut several feet from the trunk, sec-ond, an upper jump cut to remove most of the limb, and third, a collar cut at the trunk. Jump cuts prevent the limb from ripping out the bark below the cut.

PRUNING TOOLS

Hand shears are used for cuts within reach that are less than 1/2 inch in diameter.

Loppers are used for cutting limbs that are too large for hand shears and are less than 1 1/2 inches in diameter.

Bow Saws are used for larger limbs. They are very effective, easy to use, inexpensive, and available at most hardware stores or garden centers. They have replaced the old folding saw and curved pruning saw.

Pole Pruners are used for tip pruning and for thinning out limbs that are too high to reach. Most pole pruners also have a detachable saw, which is very useful.

Chain Saws need to be used when thinning out large limbs.

Index